Physical Causation

Physical Causation discusses in a systematic way an original, positive account of causation: the conserved quantities account of causal processes that Phil Dowe has been developing over the last ten years.

Dowe offers a clear and original account of causation based firmly in contemporary science. The book describes causal processes and interactions in terms of conserved quantities: a causal process is the world line of an object that possesses a conserved quantity, and a causal interaction involves the exchange of conserved quantities. Further, things that are properly called cause and effect are appropriately connected by a set of causal processes and interactions. The distinction between cause and effect is explained in terms of a new version of the fork theory: the direction of a certain kind of ordered pattern of events in the world. This particular version has the virtue that it allows for the possibility of backwards causation, and therefore of time travel.

This is an important, original book that will be widely discussed among philosophers and students working in contemporary metaphysics and philosophy of science, and among scientists with an interest in foundational issues.

Phil Dowe is Senior Lecturer in the School of Philosophy at the University of Tasmania, Australia.

Cambridge Studies in Probability, Induction, and Decision Theory

General editor: Brian Skyrms

Advisory editors: Ernest W. Adams, Ken Binmore, Jeremy Butterfield, Persi Diaconis, William L. Harper, John Harsanyi, Richard C. Jeffrey, Wolfgang Spohn, Patrick Suppes, Amos Tversky, Sandy Zabell

Other titles in the series
Ellery Eells, *Probabilistic Causality*
Richard Jeffrey, *Probability and the Art of Judgment*
Robert C. Koons, *Paradoxes of Belief and Strategic Rationality*
Cristina Bicchieri and Maria Luisa Dalla Chiara (eds.), *Knowledge, Belief, and Strategic Interactions*
Patrick Maher, *Betting on Theories*
Cristina Bicchieri, *Rationality and Coordination*
J. Howard Sobel, *Taking Chances*
Jan van Plato, *Creating Modern Probability*
Ellery Eells and Brian Skyrms (eds.), *Probability and Conditionals*
Cristina Bicchieri, Richard Jeffrey, and Brian Skyrms (eds.), *The Dynamics of Norms*
Patrick Suppes and Mario Zanotti, *Foundations of Probability with Applications*
Paul Weirich, *Equilibrium and Rationality*
Daniel Hausman, *Causal Asymmetries*
William A. Dembski, *The Design Inference*
James M. Joyce, *The Foundations of Causal Decision Theory*
Yair Guttmann, *The Concept of Probability in Statistical Physics*

Physical Causation

PHIL DOWE
University of Tasmania

CAMBRIDGE
UNIVERSITY PRESS

PUBLISHED BY THE PRESS SYNDICATE OF THE UNIVERSITY OF CAMBRIDGE
The Pitt Building, Trumpington Street, Cambridge, United Kingdom

CAMBRIDGE UNIVERSITY PRESS
The Edinburgh Building, Cambridge CB2 2RU, UK   http://www.cup.cam.ac.uk
40 West 20th Street, New York, NY 10011-4211, USA   http://www.cup.org
10 Stamford Road, Oakleigh, Melbourne 3166, Australia
Ruiz de Alarcón 13, 28014 Madrid, Spain

First published 2000

Printed in the United States of America

Typeface Times Roman 10.25/13 pt.   *System* QuarkXPress [BTS]

A catalog record for this book is available from the British Library.

Library of Congress Cataloging in Publication data
Dowe, Phil.
Physical causation / Phil Dowe.
p. cm. – (Cambridge studies in probability, induction, and decision theory)
Includes bibliographical references.
ISBN 0-521-78049-7
1. Causality (Physics)   I. Title.   II. Series.
QC6.4.C3 D69 2000
530′.01 – dc21   99-047849

ISBN   0 521 78049 7   hardback

Contents

Acknowledgements   *page* ix

1   Horses for Courses: Causation and the Task of
    Philosophy   1

2   Hume's Legacy: Regularity, Counterfactual and
    Probabilistic Theories of Causation   14

3   Transference Theories of Causation   41

4   Process Theories of Causation   62

5   The Conserved Quantity Theory   89

6   Prevention and Omission   123

7   Connecting Causes and Effects   146

8   The Direction of Causation and Backwards-in-Time
    Causation   176

References   211
Index   221

Acknowledgements

I would like to thank the very many people who have commented on and contributed to the ideas in this book, especially David Armstrong, Helen Beebee, Jossi Berkovitch, Chris Hitchcock, Peter Menzies, Hernan Miguel, the late George Molnar, Huw Price, Wes Salmon and Jim Woodward. This work has been supported by the Australian Research Council.

Permission has been granted to use material from the author's previously published articles as follows:

"Process Causality and Asymmetry," *Erkenntnis* 37 (1992): 179–196.

"Wesley Salmon's Process Theory of Causality and the Conserved Quantity Theory," *Philosophy of Science* 59 (1992): 195–216.

"Causality and Conserved Quantities: A Reply to Salmon," *Philosophy of Science* 62 (1995): 321–333.

"What's Right and What's Wrong with Transference Theories," *Erkenntnis* 42 (1995): 363–374.

"Backwards Causation and the Direction of Causal Processes," *Mind* 105 (1996): 227–248.

"The Conserved Quantity Theory of Causation and Chance Raising," *Philosophy of Science* 66 (1999): 486–501.

1

Horses for Courses

Causation and the Task of Philosophy

The aim of this book is to articulate and defend a theory of physical causation. However, before we can turn to that task there are some important metaphilosophical preliminaries. We begin in this chapter not by discussing theories of causation as such, or particular issues arising from discussions of such theories, but rather with the metaphilosophical question, *what is the task of philosophy in setting out a theory of causation?* As is the case with many philosophical questions, our question, "What is causation?" is ambiguous, and consequently the philosophy of causation legitimately involves at least two distinct tasks. In approaching the topic of causation we need to be clear about which task we are undertaking.

We begin by considering these two approaches to the task of philosophy.[1] The first is *conceptual* – to elucidate our normal concept of causation. The second is *empirical* – to discover what causation is in the objective world.[2] Clearly, it is important to be clear which task one is undertaking. In fact, we shall see that insufficient attention has been

---

1. These are not the only ways that the philosophy of causation has been conceived. Compare, for example, Skyrms's pragmatic analysis (1980) and Mellor's "middle way" (1995).
2. What is intended by these terms will become clear in the subsequent elucidation. It's the distinction Bigelow and Pargetter denote by the terms 'semantics' and 'metaphysics' respectively (1990b: 278–279), and Jackson by the terms 'conceptual analysis' and 'metaphysics' respectively (1994). In fact, it's virtually impossible to find a pair of terms that is not objectionable on some grounds. Conceptual analysis is empirical in the sense that it is concerned with how a word is used in a language, an a posteriori, empirical matter. It's also ontological insofar as it articulates the logical ontology of the concept. It's also what many philosophers mean by 'metaphysics.' On the other hand, empirical analysis is also about concepts – the concepts we think map the world; and so on.

1

paid to this metaphilosophical question, as we examine recent cases where criticism of this or that theory of causation is misdirected simply because no attention has been paid to the relevant theorist's purpose in articulating the theory.

The project of this book is to develop an empirical analysis of causation. The task of this chapter is to reflect on the nature of that project. The viewpoint adopted is that each of these 'tasks,' conceptual and empirical, is legitimate in its own right; and that it is not obvious that any particular theory of causation would adequately fulfil both tasks; and hence that we may need "horses for courses," that is, different theories for the different tasks of philosophy.

1.1 THE TWO TASKS

'Conceptual analysis' can mean a variety of things. I use it in the sense suggested by the late Oxford philosopher John Mackie in his question, "What is our present established concept of causation, of what cause and effects are, and of the nature of the relation between them?" (1985: 178). In this sense, conceptual analysis is a meaning analysis that begins with our everyday, common sense understanding of the relevant concept. That is, the way in which we commonly speak and think provides the primary data for the analysis. We look for the various intuitions that we, as mature users of the concept, can bring to bear. This is marked in philosophical discussion by judgments such as "this is counterintuitive," or "intuitively we say that . . ." Conceptual analysis is not just dictionary writing. It is concerned to spell out the logical consequences and to propose a plausible and illuminating explication of the concept. Here, logical coherence and philosophical plausibility will also count. The analysis is a priori, and if true, will be necessarily true.

Many philosophers in the twentieth century have taken the task of philosophy to be just conceptual analysis. The American philosopher Curt Ducasse conceived of the philosophy of causation in this way:

The problem of giving a 'correct' definition of the causal relation is that of making analytically explicit the meaning which the term 'cause' has in actual concrete phrases that our language intuition acknowledges as proper and typical cases of its use. (1926: 57)

Ducasse draws an analogy to a scientific hypothesis: as a scientific hypothesis aims to fit the facts ("perceptual objects and their relations"), conceptual analysis aims to fit the facts ("the intuited meanings of actual phrases in which the word to be defined occurs"). (1926: 57) There is a simple test to see whether the definition "fits the facts":

To say that a definition of ['cause'] is correct means that that definition can be substituted for the word 'cause' in any [relevant] assertion . . . in which the word occurs, *without in the least changing the meaning which the assertion is felt to have.* (1926: 57)

Hart and Honore take a similar view:

The ordinary man has a quite adequate mastery of various concepts within the field of their day-to-day use, but, along with this practical mastery goes a need for the explicit statement and clarification of the principles involved in the use of these concepts. (1985: 26)

Conceptual analysis as just described is not *revisionist*. In these cases there is no intention to improve or replace the established concept of cause; the intention is merely to explicate the concept as given. Conceptual analysis, in the sense discussed so far, is an *explication* of everyday concepts. Sometimes, however, conceptual analysis may be revisionist. Such an approach will involve proposing changes to the way we currently think, perhaps on the grounds of logical consistency or economy.[3]

On the other hand, empirical analysis seeks to establish what causation in fact *is* in the actual world. Empirical analysis aims to map the objective world, not our concepts. Such an analysis can only proceed a posteriori.

This program has variously been called "empirical metaphysics" (David Armstrong), "ontological metaphysics" (Aronson 1982), "speculative cosmology" (Jackson 1994), "physicalist analysis" (Fair 1979: 233) and "factual analysis" (Mackie 1985: 178).

According to Mackie, this is the way David Hume's famous and influential regularity theory is to be understood:

In his definitions, which aim at reform rather than analysis of our ordinary concepts, [Hume] equates causation as it really exists in the objects with regular succession. (1974: 59)

Mackie himself wishes to understand the ultimate task of philosophy in this way:

3. Compare Strawson's distinction between revisionary and descriptive metaphysics (1959). By 'metaphysics' Strawson means what I mean by 'conceptual analysis.' See also the 'reformatory analysis' of Ackerman (1995).

The causation I want to know more about is a very general feature . . . of the way the world works: it is not merely, as Hume says, *to us*, but also *in fact* the cement of the universe. (1974: 2)

[T]his is an ontological question, a question about how the world goes. In Hume's phrase, the central problem is that of causation 'in the objects.' (1974: 1)

Other philosophers who have adopted this general approach to causation include David Fair: "the hypothesised relationship between causation and energy-momentum flows is expected to have the logical status of an empirically discovered identity" (1979: 231); Jerrold Aronson: "this 'ontological' approach . . . allows us to perceive causation in objective terms, retaining its rightful place in the physical sciences" (1982: 291); John Bigelow and Robert Pargetter: "causation is a robust ingredient within the world itself" (1990b: 294); and perhaps Dorothy Emmet: "[we need] to get beyond the discussion of logic and epistemology of causal statements and get into the ontology underlying them" (1985: 5). All these philosophers agree that to understand causation we need to go beyond words, to look at the world. This is what is intended by the label 'empirical analysis.'

But 'empirical analysis' as discussed so far covers several distinct concepts of analysis. For example, a given analysis may have the status either of a contingent truth, or of a necessary truth in the style of Kripke (1980). Wesley Salmon (1984) seeks merely to articulate what causation is as a contingent fact, while others such as Bigelow and Pargetter (1990b) try to establish what causation is as an a posteriori necessity. We will use the term 'empirical analysis' to cover the first of these options. In this sense, the account of causation developed in this book is an empirical analysis.

Empirical analysis also may or may not be revisionist. A theory of causation that is empirically true may or may not be a good account of our common sense understanding of causation. Whether it is or not is itself an empirical matter. Hume, for example, thought not: one plausible reading of Hume attributes to him two theories of causation – the conceptual analysis of the everyday notion in terms of necessary connections or hidden power, and the empirical analysis in terms of constant conjunction,[4] together with the advice that we should free ourselves of the false connotations of the common sense notion. If

4. See Chapter 2, section 2.1; see also Beauchamp and Rosenberg (1981), especially Chapter 8.

4

this is correct, then Hume was a revisionist. On the other hand, a nonrevisionist empirical analyst would claim either that empirical analysis and conceptual analysis coincide, so that common sense delivers the empirically correct theory, or that for some reason common sense need not match the empirically correct theory where there is a discrepancy.

I.2 CONFUSING THE TASKS

Insufficient attention to this distinction has been the source of numerous errors in the literature. These mistakes typically arise when philosophers criticise a theory of causation without paying attention to what type of theory it is intended to be. To criticise a theory that aims to explicate the meaning of everyday usage on the grounds of what quantum physics tells us about reality, for example, or to criticise a theory that claims only to provide an empirical analysis on the grounds of its shortcomings as a meaning analysis is to seriously misdirect the criticism. Yet it seems that very commonly this is what is done.

One very common form of this error is to criticise empirical theories on the grounds that they do not adequately account for this or that feature of the way we talk about causation.[5] For example, Beauchamp and Rosenberg claim that many critics of Hume and J. S. Mill commit this error, by wrongly assuming that their theories are trying to account for everyday usage. Collingwood, Flew and Anscombe are all implicated (1981: 285–286). Thus, Mill and Hume "ought not to be faulted for neglecting to provide the analyses they never intended to provide and had no philosophical reason to undertake" (1981: 294).

Although not mentioned by Beauchamp and Rosenberg, Ducasse is most explicit in promoting this kind of criticism of Hume:

I believe [Hume's] account of the nature of causation – simply as *de facto* succession – represents an incorrect analysis of the ordinary notion of cause . . . To make evident the incorrectness of that analysis it will be sufficient to show, on the one hand, that there are cases which conform to Hume's definition but where we judge the events concerned not to be related as cause to effect; and on the other hand, that there are cases which do not conform to Hume's definition but which we nevertheless judge to be cases of causation. (1976: 69)

5. Bigelow and Pargetter warn, "It will be no objection to our proposals to cite one or another causal idiom which we have failed to explain" (1990b: 278). See also Millikan (1989) and Neander (1991).

Ducasse then proceeds to provide cases of both types: the first type is a case such as Reid's example of day following night (regular succession but not causation) and the second type is illustrated by Ducasse's brown paper parcel wrapped in string, which on being pressed at one end glows at the other (causation without regular succession).[6]

A more recent example is the Dutch physicist Dieks's criticism of the transference theory of Fair (1981). Dieks, while recognising that Fair is attempting to discover "the true nature of the causal bond" as an ontological category, nevertheless concludes his critique: "So we see that the new analysis of causation has its own share of problems, that is, divergences from the everyday language use of the concept 'cause'" (Dieks 1981: 105).

Perhaps these critics desire in a theory of causation some kind of conceptual analysis of the everyday concept, and they may have good reasons for such a preference. But this debate ought to be conducted explicitly at the metaphilosophical level to avoid the possibility of errors of the type that I have just described.

Another common error is to require that an empirical analysis hold good for all logically possible worlds. As we have seen, not all theorists are attempting to provide such an analysis. This mistake is made by John Earman, who uses a possible-worlds argument against Aronson, when Aronson's theory of causation as energy/momentum transfer is intended as an empirical analysis which holds in the actual world.[7] Earman's objection that we would call a collision in a possible world where energy is not conserved 'causal' (1976: 24) ignores the fact that Aronson is not seeking to provide a necessary identity. The same mistake is made by Michael Tooley (1987: chap. 7), who uses a range of possible-world arguments to disprove Salmon's theory of causation (Salmon 1984),[8] when Salmon's explicit purpose is to discuss causation as it is found in the actual world. Again, the mistake arises

6. A similar example of this kind of mistake is Hart and Honore (1985: 22, 34), who criticise Mill on the grounds that his theory neglects several aspects of normal speech; and Hugh Mellor, who argues that both Hume and Salmon fail in their respective theories because they do not capture all the connotations of causation (1988: 231), when neither Hume nor Salmon are attempting any sort of conceptual analysis of the everyday meaning of 'cause.'

7. "The transference theory is intended to make sense of how causation takes place in *this* world, . . . not in some alien universe where the laws of physics do not in the least resemble ours" (Aronson 1982: 302).

8. For a detailed analysis see Dowe (1989).

because insufficient attention is paid to the author's metaphilosophical intentions.

Other examples could be given, but this suffices to illustrate the point: there is more than one distinct task a theory of causation might be asked to do, and it is essential to understand what the intentions of the author are with respect to that task before criticising the theory. This is not to say that a critic is obliged to agree with an author about the task of philosophy, but disagreements about the task of philosophy need to be distinguished from disagreements about whether a theory fits the task for which it was designed.

I.3 THE LEGITIMACY OF EMPIRICAL ANALYSIS

It is uncontroversial that conceptual analysis has a legitimate role to play in philosophy. On the other hand, the legitimacy of an empirical analysis of causation *has*, at times, been questioned. We shall briefly consider two lines of criticism.

The first argument is one that has sometimes been used in the defence of understanding philosophy purely as linguistic analysis. According to this argument it is not possible to know anything about any language-independent entity called 'causation,' because we have no procedures for investigating such an entity. (See, for example, Alston 1967: 388.) However, the empirical analyst can reply that there are procedures for investigating such an entity, namely, the methods of science, which is in the business of investigating language-independent objects. Empirical philosophy can draw on the results of science, and so can investigate such concepts, in this case causation 'in the objects.'

For example, science has shed light on the nature of energy. 'Energy' has today a technical scientific meaning. When asked the meaning of the term, we simply give the scientific definition. Adequate explication of that definition took several centuries, but prior to that achievement, the term simply had a vague range of meaning in everyday language, somewhat as the word 'cause' does today. We can say that application of the scientific method of theorising and experimentation produced an 'empirical analysis' of energy. In the same way, science may reasonably be expected to throw light on the language-independent entity called 'causation.'

The second argument is related to the first. This argument asserts that it is not the role of philosophy to deal in synthetic a posteriori matters, which is the exclusive task of science. Ducasse, for example, held this view: "No discovery in any of the sciences has or ever can have any logical bearing upon the problems of philosophy" (1969: 120).

The most direct way to answer this argument is to show how science does inform philosophy about causation, which would indicate that fruitful interaction is possible. Data of this sort is not difficult to find: quantum mechanics has (arguably) shown us that the law of causation (interpreted substantially)[9] is false; entropy and the time reversibility of the basic laws of nature inform us about causal asymmetry; tachyons, Bell's Theorem, and kaon decay have alerted us to the logical possibility of backwards-in-time causation; biomedical science and econometrics have shown us how to directly test causal claims via path analysis; attribution theory in cognitive psychology tests for us other aspects of causal theorising; and so on. This indicates that science is able to inform philosophy, and that empirical analysis in philosophy can draw on scientific results – on empirical, synthetic facts – and use them in analysis. Of course, this might not convince someone like Ducasse, whose pronouncement was most likely prescriptive with respect to the nature of philosophy. But if we have clear cases of such an enterprise, then the burden of proof lies with those who deny the possibility (see Dowe 1997a: sec. 5).

The reply to this second objection throws light on the nature of empirical analysis. There are many ways that science does inform philosophy; and where philosophy takes these results into account, *that* is empirical analysis. So we may reasonably claim to be warranted in assuming that the task of empirical analysis is legitimate.[10]

This, of course, is not to deny that conceptual analysis is legitimate. In undertaking an empirical analysis, as we will in this book, no implication is intended that this is the only way to do philosophy.

It may be objected here that although it is legitimate in its own right, empirical analysis cannot be undertaken without conceptual analysis. Even revisionist approaches require some degree of conceptual analy-

---

9. That is, 'every event has a sufficient cause.' This informs our thinking about the concept of causation if, for example, we are accustomed to thinking that causes are sufficient conditions for their effects, and yet are forced to accept that there are cases that we cannot but call "causation," where the full cause is not a sufficient condition for its effect.
10. In this book I do not offer a full-scale case for the legitimacy of empirical analysis qua empirical analysis. However, the brief arguments just given should serve as adequate preliminaries to the task undertaken in this book – namely, to explicate an empirical theory of causation. For a defence of an alternative view, see Tooley (1987; 1990).

sis of the common usage, or else it would not be clear in what way or to what degree the theory *is* revisionist, or even whether it deserves the name 'cause.' To do this one must have some account of common usage. As David Lewis says,

Arbiters of fashion proclaim that analysis is out of date. Yet without it, I see no possible way to establish that any feature of the world does or does not deserve a name drawn from our traditional . . . vocabulary. (1994: 415)

So, it will be objected, even in undertaking empirical analysis, some attention to conceptual analysis is required. We want to know the true nature of the thing *we call* causation rather than the true nature of something altogether different. Thus conceptual analysis is needed at the very least to serve as a 'rough guide' or an 'introduction' to a empirical analysis (compare Mackie 1974: 2).[11]

Bigelow and Pargetter address (or perhaps, avoid) this issue (1990a; 1990b) by taking the concept of cause as a primitive notion so far as meaning analysis goes. They comment,

It is important to recognise that there is a bridgeable but problematic swamp lying between the metaphysics and the semantics of causation. And in offering a metaphysics of causation, we are not pretending to solve all the semantic problems . . . As far as semantics is concerned, this causal relation is primitive . . . Our task is metaphysical, not semantic. (1990a: 102; 1990b: 278–279)

Unfortunately, this does nothing to answer the objection that we need to know whether the thing we find in science deserves the name 'cause.'

However, I think this objection can be met, as follows. In drawing explicitly on scientific judgements rather than on intuitions about how we use the word, we nevertheless automatically connect to our everyday concept to some extent, because the word 'cause' as scientists use it in those scientific situations must make some 'historical' or 'genealogical' connection to everyday language. This is especially likely given on the one hand that 'cause' is not a technically defined term in any scientific theory, and on the other hand that the word is not being used playfully or ironically (as is the word 'quark' or the phrase 'eight-fold way'). So to deny that there is an adequate connection is to deny that scientists are competent users of English.

---

11. Jim Woodward pointed out to me that if objectivity is part of our concept of causation, then an adequate conceptual analysis will need to respect empirical results. See also the approach of Mellor (1995).

The connection may be to some extent tenuous, in the sense that the resulting analysis is highly revisionist. We certainly do want to avoid assuming a priori that for any feature X of our everyday concept of causation, causation actually has feature X. Nevertheless, the historical connection is sufficient to warrant our use of the term.

To return to our previous example, before the development of classical physics 'energy' was a word from everyday vocabulary that, over centuries of scientific endeavour, came to have a very precise scientific meaning. But at no stage of this development was it felt necessary to spell out how the emerging scientific concept differed from the everyday concept, or to what extent it differed. To some extent what one then said was a matter of the conventional assignment of labels: one can imagine comments such as "No, what you're talking about is really 'force' in the scientific sense, not energy." Yet also in some ways the whole development showed that in certain ways common sense was mistaken, for example, in the way it had incorporated Aristotelian physics. However, I note again that it is not my concern to establish the extent to which common sense is wrong. But the use of the term 'energy' bore an historical connection to the scientific definition, sufficient to warrant that use of the term, even though popular usage of the word continued to be somewhat vague and loose. So the connection between our everyday concept 'cause' and the result of a satisfactory empirical analysis is guaranteed just by virtue of the fact that scientific discussions employ the English word 'cause.'

A further objection may be raised here, as follows. The distinction between conceptual analysis and empirical analysis is not as cut-and-dried as has been presented. Any empirical analysis will still be a kind of conceptual analysis, for example, of scientists' usage of the word. This is not the same as analysing the concept in everyday thought and language, but it is the same type of activity.

However, in replying to this objection, we need to note immediately that the task of empirical analysis as undertaken in this book is not a conceptual analysis of scientists' usage of a term. It is an attempt to understand causation in the world. Certainly we need to look to science to provide us our best information about the world, but for a word such as 'cause' – which is not a technical term in science – scientists' usage may reflect aspects of the everyday concept that are not part of the concept as it emerges from science itself, or that even contradict that concept. In his book *Time's Arrow and Archimedes' Point* (1996b), Huw Price accuses scientists of doing this with respect to the direction of time. According to Price, science itself is time-symmetric, and the direction of causation is something that we impose on the world. But scientists are not always immune from slipping back into thinking and speaking as if there were an objective direction, and Price catalogues numerous significant examples of what he calls "that old double standard." While I don't entirely agree with Price's views (see Chapter 8), this example does illustrate the possibility that scientists' linguistic habits may not be the best guide to the structure of a concept emerging from science. Empirical analysis looks to science itself rather than to the linguistic practices of scientists.

Does this undermine my claim that scientists' use of the word 'cause' is sufficient to establish that an empirical analysis of 'cause' has a right to the word? Not at all. Again we must urge that no assumption can be made about the extent to which the common use of the term will match the empirical analysis. The existence of discrepancies is not in itself reason to deny the right. That there are not too many discrepancies for the analysis to have the right is guaranteed by the scientists' use of the term, together with the assumption that scientists are competent users of the language.

The objection may be modified as follows. The distinction between conceptual analysis and empirical analysis is not so cut-and-dried; any empirical analysis will still be a kind of conceptual analysis, for example, of the concept implicit in scientific theories. I am happy to grant this, and accept that in this sense the distinction is not as cut-and-dried as I have presented it. As a scientific realist (see Dowe 1996), I take it that scientific theories are the best guide to the structure of reality, and therefore that an empirical analysis, which seeks an analysis of an aspect of the structure of reality, must look to scientific theories. So I am happy to think of the task of empirical analysis as a conceptual analysis of a concept inherent in scientific theories. Further, the issue of scientific realism makes no practical difference to the work in this book, since someone who denies such a connection between theory and reality may be happy nonetheless with the idea of empirical analysis as analysis of the concept inherent in scientific theories.

I.4 HORSES FOR COURSES

If both conceptual analysis and empirical analysis are legitimate tasks, is it likely that one theory of causation can fulfil both tasks? Is a successful conceptual analysis likely to be a successful empirical analysis? These questions can only be properly answered a posteriori, that is, by looking at actual theories and seeing if they do adequately fulfil both tasks. Here it will suffice to point out that we cannot assume that the best conceptual analysis is also the best empirical analysis, or vice versa.

If this is right, then we probably need horses for courses. An empirical theory should not be taken as providing a meaning analysis of common sense; and a conceptual analysis should not be read as an account of the true nature of causation in the world. In the case that we are interested in both questions we may need two horses: a theory for each task. For example, it may be that of the various contemporary theories of causation the manipulability or counterfactual theories best address the conceptual question, and that regularity or transference theories best address the empirical question. There are many possibilities here, of course,[12] but we can conclude from the above discussion that one may well need a stable of more than one horse in order to negotiate adequately contemporary philosophical discussion about causation.

This book presents an empirical analysis of causation; no meaning analysis is undertaken. The approach taken here is that the empirical theory can be articulated without looking closely at the everyday concept 'causation.' The starting point will instead be hints taken from science – for example, the idea of a causal process and the distinction between causal and noncausal processes found in special relativity; the emergence of probabilistic causality in biomedical science; and the backwards-in-time causation postulated in the transactional account of quantum interactions. These cases are employed in the assessment of various empirical theories of causation. For this reason it is appropriate to call this a 'physical' theory of causation.

The outline of the book is simple. In Chapters 2–4 we consider some of the major theories of physical causation: David Hume's regularity theory, David Lewis's counterfactual theory, Patrick Suppes's probabilistic account, the transference theory of Jerrold Aronson and David Fair, and the process theory due to Wesley Salmon. In Chapters 5–8

---

12. For example, there are other approaches to the connection between conceptual and empirical analysis, such as that of Menzies (1996), who utilises the Lewis-style approach to analysis that gives a functionalist account of the meaning, to which is added an identity theory, as we will see in Chapter 2.

the positive account – the Conserved Quantity Theory – is articulated and defended. In Chapter 5 an account of causal processes and interactions is presented. Chapter 6 deals with the special case of preventions and omissions. In Chapter 7 the notion of a causal process is used to explicate the connection between causes and effects. And finally, in Chapter 8, an account is offered of the direction of causation.

Having done this, one could examine the empirical analysis to see to what extent it does serve as a conceptual analysis of common sense meaning. However, such an examination is largely beyond the scope of this book. Of course, there is no a priori claim about the adequacy of the empirical analysis qua conceptual analysis. Given the complexity of the issues, we should be happy if we can get one horse out onto the track!

2

Hume's Legacy

Regularity, Counterfactual and Probabilistic Theories of Causation

Our examination of causation begins with the work of David Hume, whose regularity theory is perhaps the best-known philosophical theory of causation. Modern versions of that theory include the counterfactual account due to Lewis and the probabilistic theory due to Suppes. However, there are problems with these regularity accounts. First, the Humean deterministic accounts are rejected on the grounds that science yields examples of indeterministic causation; and second, the probabilistic accounts of causation, including Lewis's counterfactual probabilistic theory, are shown to fall to a well-directed example of chance-lowering causality. This paves the way for the examination of a putatively non-Humean theory – the transference theory – in Chapter 3.

II.1 HUME'S REGULARITY ACCOUNT

The usual starting place for contemporary discussions of causation is the work of David Hume. There are numerous reasons for this, even though his work is well over two hundred years old. Over that period Hume's analysis has stood as the foundation of empiricist work on causality, and even today his actual account remains a "live option" (for example, Beauchamp and Rosenberg 1981). Further, his views still delineate much of the field, with many of the issues involved in modern discussions arising from Hume's insights. For these reasons, it is a good place for the present investigation to begin.

It is possible to find in Hume's account both a conceptual analysis of our everyday notion of causation, and an empirical analysis of causation as it is "in the objects." Since Hume offers at least two further distinct theories, which won't be discussed here, it is necessary to begin with a comment about *which* two theories are under consideration.

Hume's conceptual analysis of our everyday concept is presented by Hume in order to be rejected. Here Hume outlines what we think we mean by 'causation' and related terms such as 'necessary connection' and 'causal power,' and the main conclusion of his enquiry is that such concepts are not meaningful. Hume's empirical analysis is his famous "constant conjunction," or regularity theory. The key claim of this account is that for two events to be cause and effect, they must instantiate a universal sequence; so that all occurrences of similar causes are followed by similar effects. These two theories will be examined in detail here. In addition to these, one can also discern a so-called counterfactual theory, and also a "natural" theory, wherein Hume suggests that a cause is an object whose idea determines the mind to form the idea of its effect (Hume 1975: 75–76). The former will be considered in a later section, the latter is beyond the scope of the present discussion.[1]

II.1.1 Hume's Conceptual Analysis

Hume's discussion of causation has a negative and a positive component. The negative component, the rejection of our ordinary concept as meaningless, forms the major part of his account in both of his two books, the *Treatise*[2] and the *Enquiry*.[3] In brief, the negative argument runs as follows:

Meaningful concepts derive from sense impressions. The impression from which our idea of causal power or necessary connection derives cannot be discovered in any of the instances commonly appealed to by philosophers. To find such an impression one would have to have an experience of the cause, on a single occasion, without reference to what happens to similar objects on other occasions, and from that experience alone be able infallibly to infer what will follow as the effect.

1. Usually discussions of Hume's "two theories" concern his regularity theory and his natural theory. See, for example, Garrett (1993).
2. *A Treatise of Human Nature.* All references to the *Treatise* will be to the page numbers of the second edition of Selby-Bigge and Nidditch (1978).
3. *An Enquiry Concerning Human Understanding.* All references to the *Enquiry* will be to the page numbers of the thirrd edition of Selby-Bigge and Nidditch (1975).

Such an impression cannot be discovered; therefore, the concept is meaningless.[4]

However, in order for this negative argument to work, Hume does employ an account of our everyday notion of causation. In a section entitled "Of Probability, and of the Idea of Cause and Effect" in Part 3 of Book 1 of the *Treatise*, Hume begins his discussion of causation with these words:

> To begin regularly, we must consider the idea of *causation*, and see from what origin it is deriv'd. . . . Let us therefore cast our eye on any two objects, which we call cause and effect, and turn them on all sides, in order to find that impression, which produces an idea of such prodigious consequence. (*Treatise*: 74–75)

Notice that Hume says "which we call." He is not committing himself at this stage to saying anything about what causation really is. Thus, having decided that the idea of causation must be derived from some relation among objects rather than from particular qualities of objects, and that the relation of *contiguity*[5] must be part of the account, Hume concludes,

> We may therefore consider the relation of CONTIGUITY as essential to that of causation; at least may suppose it such, according to the general opinion, till we can find a more proper occasion to clear up this matter. (*Treatise*: 75)

Thus Hume is not claiming at this stage that causation as it really is in the objects involves this relation of contiguity, but just that the idea of causation according to the general opinion involves contiguity. He is leaving open the former question.

Having added to the account the relation of *priority*,[6] Hume finds that he can discover no further feature in his impressions of a single-case of causation. But this is not satisfactory, Hume feels:

> Shall we then rest contented with these two relations of contiguity and succession, as affording a compleat idea of causation? By no means. An object may be contiguous and prior to another, without being consider'd as its cause. There is a NECESSARY CONNEXION to be taken into consideration; and that relation is of much greater importance, than any of the other two above-mention'd. (*Treatise*: 77)

---

4. This argument is given both in the *Treatise* and in the *Enquiry*; although the version in the *Enquiry* is much streamlined (1975: chap. 7).
5. Objects are *contiguous* if they are immediately adjacent in space and time.
6. The thesis that causes are always prior in time to their effects.

Notice the significance of the phrase "without being consider'd as." Hume is analysing what *we take* to be the concept of causation. Now, several chapters later, Hume finds that since there is no impression from which this "common" relation is derived, it is therefore meaningless:

Thus upon the whole we may infer, that . . . when we speak of a necessary connection betwixt objects, and suppose, that this connection depends upon an efficacy or energy, with which any of these objects are endow'd; in all these expressions, *so apply'd*, we have really no distinct meaning, and make use only of common words, without any clear and determinate ideas. (*Treatise*: 162)

In other words, our everyday notion of causation is meaningless and needs revision. But Hume's task at this earlier stage (*Treatise*: sec. 1.3.2) of the discussion is simply to outline the details of the notion as it is commonly used and understood.

Thus we can say that Hume offers an account of the common sense concept, which can be summarised as follows.

HCA:[7] Object A is the cause of object B iff

1. A and B are contiguous or linked by a chain of contiguous events;[8]
2. A precedes B; and
3. A necessitates B.

Other terms that are roughly synonymous, such as 'efficacy' and 'causal power', can be defined in similar ways.[9]

Why does Hume offer such an analysis, given that it turns out to be meaningless? Because the analysis is necessary first to identify which aspect of the concept is problematic (viz. Part 3), and second to serve as a basis of the proof that the concept is meaningless. How the analysis serves as a basis of the proof can be illustrated by the following sample of Hume's reasoning:

Now nothing is more evident, than that the human mind cannot form such an idea of two objects, as to conceive any connection betwixt them, or comprehend distinctly that power or efficacy, by which they are united. Such a con-

---

7. For Hume's Conceptual Analysis, in keeping with the terminology of the previous chapter.
8. "Tho' distant objects may sometimes seem productive of each other, they are commonly found upon examination to be link'd by a chain of causes, which are contiguous among themselves, and to the distant objects" (*Treatise*: 75).
9. See Garrett's discussion (1993: 178).

nection wou'd amount to a demonstration, and wou'd imply the absolute impossibility for the one object not to follow, or to be conceiv'd not to follow upon the other: Which kind of connection has already been rejected in all cases. (*Treatise*: 161–162)

For Hume to demand that the illusive impression be such that it implies the "absolute impossibility for the one object not to follow," it is necessary that the concept essentially involve requirement (3) as stated in HCA. Without this requirement Hume's argument would not have been valid. Thus, given that the negative component of his discussion is the major component, the analysis labeled HCA is very important to Hume's argument.

II.1.2 Hume's Empirical Analysis

Having ruled out necessary connections between cause and effect and any "hidden power" of cause to produce effect, Hume turns to his positive account. The most we can find in a causal relation, according to Hume, is 'constant conjunction' or regularity of sequence. In other words, a cause is, in fact, always followed by its effect, but beyond this there is no necessity. This account, for which Hume is most famous, is usually called the *regularity* theory. We shall now consider a brief outline of Hume's regularity theory of causality with special reference to some of the issues picked up in this and subsequent chapters, ignoring exegetical debate, and taking the orthodox reading of Hume.

In the *Treatise* Hume presents this definition of a cause:

An object precedent and contiguous to another and where all the objects resembling the former are placed in a like relation of priority and contiguity to those objects that resemble the latter. (*Treatise*: 169, Hume's italics)

The resemblance condition "placed in a like relation" requires that objects may be sorted into types, or kinds, and that for something to be a cause, all objects of that kind must be followed by a similar effect. Regularity accounts take it that every causal sequence instantiates a universal regularity. The word 'contiguous,' which means that causes must be in contact with their effects, rules out by definition any action at a distance. The word 'priority' indicates that causes must precede their effects. So causation is reduced to resemblance (a relation of ideas) and the spatiotemporal relations of priority and contiguity (matters of fact). This regularity theory can be summarised as follows:

HEA:[10] Object A is the cause of object B iff

1. A and B are contiguous or linked by a chain of contiguous events;
2. A precedes B; and
3. every A is followed contiguously by a B.

Hume intends this to be an empirical analysis,[11] in the sense of the previous chapter. It is intended to explicate causation as it really obtains in the world, independent of our thinking. For example, Hume says,

> As to what may be said, that the operations of nature are independent of our thought and reasoning, I allow it; and accordingly have observ'd, that objects bear to each other the relations of contiguity and succession; that like objects may be observ'd in several instances to have like relations; and that all this is independent of, and antecedent to the operations of the understanding. (*Treatise*: 168–169)

There are three points about HEA that bear special relevance to the present work. First, the word 'priority' is significant. Hume thereby defines 'cause' in terms of 'time,' presuming an already-established temporal order. We may call this the *temporal* theory of causal direction. This entails that the causal relation is *asymmetric*.[12] In fact, one advantage of the temporal theory is that it provides a ready-made explanation of causal asymmetry, and of the distinction we recognise between cause and effect. But it also rules out the possibility of simultaneous and backwards-in-time causation. Because of this implication, there are reasons found in modern science for avoiding the temporal theory, as we shall see in Chapter 8.

Second, Hume has conjoined a condition concerning a particular instance of causation, with a condition concerning the class of all such cases: "*an* object precedent and contiguous to another, where *all* the objects resembling the former are placed in like relation" (*Treatise*: 169; my italics). Hume is taking causation to be primarily a relation between particular events, but is asserting what would today be called a *supervenience* thesis[13] about those causal relations. In general, Hume's super-

---

10. For Hume's Empirical Analysis.
11. A good case for this is made in Mackie (1974: chap. 1).
12. That is, if A causes B, then it is not the case that B causes A.
13. A simple way to think of supervenience is: A supervenes on B iff there can be no difference in A without a difference in B.

venience thesis is that all facts supervene on actual particular matters of fact; in other words, causal facts (say) are logically determined by facts about actual particular matters of fact. In the case of causation, this involves two supervenience theses. The first is that the causal relations supervene on causal laws; that is, a causal relation holds by virtue of a certain causal law holding. The second is that causal laws (and more generally all laws, if there are noncausal laws of nature) supervene on actual particular matters of fact; that is, the truth of a law is determined by the obtaining of a certain pattern of actual particular matters of fact.[14]

Third, and most significant for the argument of this chapter, the occurrence of the word 'all' indicates a connection between causality and determinism: that, in one sense of 'sufficient,' a cause is a *sufficient condition* for its effect. That is, if an event is to be called a 'cause,' then its occurrence must be a sufficient condition for another event, the effect. Here, 'sufficient condition' is used in the following actualist sense:

C is a *sufficient condition* for E if and only if all Cs
are followed by Es.

The 'all' here contains an implicit reference to the actual world: it means if and only if every C that occurs in the actual entire spacetime universe is in fact followed by an E, then C is a sufficient condition for E.[15] This actualist sufficient condition can be expressed as a probability relation (see Suppes 1970: 34),[16] where the probabilities are construed as actual frequencies:

$$P(E|C) = 1$$

14. For a discussion of Hume's supervenience theses, see Tooley (1984; 1987: chap. 6).
15. This is to be distinguished from modal senses of sufficient, where a sufficient cause necessitates its effect (see Mellor 1995: chaps. 1, 2). This distinction between different senses of 'sufficient,' and also the correlative distinction between modal and actualist senses of 'necessary condition,' are not always clearly recognised by philosophers. For example, in his influential introductory text John Hospers at one point introduces an actualist sense of necessary condition: "A is a necessary condition for B when, in the absence of A, B never occurs" (1990: 202). A few pages later he slips over to a modal sense: "When we say that C is a necessary condition for E we mean only that if C had not occurred E would not have occurred" (1990: 208).
16. Suppes, while not one who takes probability to mean frequency, nevertheless is using 'sufficient' in an actualist sense.

This is an expression of the maxim "every effect has a sufficient cause," which asserts the deterministic character of causality. It is not possible, on Hume's account, for causes to be less than deterministic. *Indeterministic causality* is ruled out as a conceptual possibility. (This must not to be confused with the so-called law of causality, or universal determinism, which makes the stronger claim that every event has a sufficient cause.)[17] Thus the actualist sufficient condition implicit in HEA implies just one aspect of the doctrine of determinism – that a cause is always followed by its effect. This feature of Hume's account is addressed by the argument presented in the following section.

II.1.3 Rival Interpretations

Before we turn to that argument, however, it is worth pointing out that the interpretation just given of Hume's intentions makes much more straightforward sense of his total argument than do its common rivals. One major rival takes Hume to be offering his regularity theory as a conceptual analysis of our concept of causation.[18] Amongst other things, this makes no sense of certain of Hume's statements, such as "There is a NECESSARY CONNEXION to be taken into consideration; and that relation is of much greater importance, than any of the other two above-mention'd" (*Treatise*: 77). Another, more recent, rival interpretation is the so-called revisionist interpretation due to John Wright (1983), Edward Craig (1987) and Galen Strawson (1989),[19] according to which Hume believed that there are powers in nature, but that we can't know anything about them. On this account Hume's empirical analysis is something like HCA, while HEA delineates our epistemic limits. This makes sense of the quote just given, but makes no sense of the later stages of Hume's argument, where he speaks as if HEA concerns causation "in the objects,"[20] or of the claims that our idea of necessary connection and causal power in the objects is in fact

17. There is a related causal principle that also follows from HEA, namely, the principle of causal regularity, "same cause, same effect." In Hume's words, "The same cause always produces the same effect" (1975: 70). This principle will not concern us in this chapter.
18. See, for example, Ducasse (1976: 69), and the discussion on that point in the previous chapter.
19. See also Blackburn (1990), Broughton (1987), Costa (1989), Pears (1990) and Winkler (1991).
20. For example, see the quotation given earlier from the *Treatise*: 168–169.

meaningless.[21] Strawson attempts to explain the latter discrepancy by the theory that Hume is not serious about concept empiricism, but is just paying lip service to the empiricist tradition by starting his discussion with a statement of that thesis (1989). But despite Strawson's lively attempts to defend it, this claim lacks credibility, not least because it's not until the end of the argument that Hume actually comes out with the conclusion that those notions are meaningless; and at the very *end* of his discussion in the *Enquiry,* Hume summarises the entire argument explicitly as an argument showing from concept empiricism the meaninglessness of the idea of causal power or necessary connection in the objects:

> To recapitulate, therefore, the reasonings of this section: Every idea is copied from some preceding impression or sentiment; and where we cannot find any impression, we may be certain that there is no idea. In all single instances of the operation of bodies or minds, there is nothing that produces any impression, nor consequently can suggest any idea, of power or necessary connection. (*Enquiry*: 78)

As opposed to these alternative interpretations of Hume's intentions, the account sketched here makes sense of both sets of quotes, as I have argued.[22]

II.2 INDETERMINISTIC CAUSATION

A significant step forward in philosophy of science over the last twenty years[23] has been the emergence of probabilistic causality as a serious empiricist position. This, of course, has required the recognition that determinism and causality are not necessary partners. It is now possible, from the perspective of modern science, to refute Hume's requirement that a cause be a sufficient condition for its effect.

Suppose that we have an unstable lead atom, say Pb^{210}. Such an atom may decay, without outside interference, by α-decay into the mercury atom Hg^{206}. Suppose the probability that the atom will decay in the next minute is x. Then

$$P(E|C) = x$$

21. For example, see the quotation given earlier from the *Treatise*: 162.
22. On the other hand, Strawson's big positive argument is that Hume could not really believe the real world in itself is just a collection of particular spatiotemporal matters of facts. I make no attempt to absolve Hume on this score.
23. Possibly due originally to Reichenbach.

where C is the existence of the lead atom at a certain time t_1, and E is the production of the mercury atom within the minute immediately following t_1. In fact x is a very small number, since the half-life of Pb^{210} is about 7×10^8 seconds (Enge 1966: 225). Now consider the following argument:

1. The connection between C and E is a causal connection.
2. The probability $P(E|C) = x$ is irreducible.
3. Therefore, not all causes are sufficient conditions for their effects.

A powerful case can be made for accepting each of the premises of this argument. Let's start with the first premise. In order to rattle everyday intuitions, consider the following argument.

If I bring a bucket of Pb^{210} into the room, and you get radiation sickness, then doubtless I am responsible for your ailment. But in this type of case, I cannot be morally responsible for an action for which I am not causally responsible. Now the causal chain linking my action and your sickness involves a connection constituted by numerous connections like the one just described. Thus the insistence that C does not cause E on the grounds that there's no deterministic link entails that I am not morally responsible for your sickness. Which is sick.

It may be objected here that I am making an illicit appeal to everyday intuitions about meaning – illicit since I am using them to criticise Hume's empirical theory and since I myself am seeking to develop an empirical theory. The latter is true. However, this does not preclude us from utilising such considerations for the purpose of shaking preconceptions, such as causal determinism, if those preconceptions have no scientific basis. I don't think this is an illicit strategy.

But I also have a second argument. This is simply to point out that scientists describe such cases of decay as instances of *production*[24] of Hg^{206}. Now 'production' is a near-synonym for 'causation' (Mellor 1988: 231), so it's not just folk engaged in everyday discourse, but also scientists at work, who pronounce C to be a cause of E. It may be objected here that we should be appealing not to what scientists say at offhand moments, but to the concepts inherent in scientific theory and explanation. I agree that such considerations carry more weight. But scientists' language does illustrate the fact that they see no scientific obsta-

24. For example, Arya (1974: 386, 389, 397), and see the word 'cause' (1974: 402–405); and Enge (1966: 227–228).

cle to applying our everyday concept of causation to indeterministic cases. If we were to examine the use of causal concepts within actual theories (a task that will be addressed in Chapter 5), we would find, for example, that the causality concept involved in special relativity presents no obstacle to taking indeterministic processes to be causal.

This is not a knock-down argument for premise (1) (such arguments are very rare in philosophy). It is always open to the causal determinist to deny that any of these cases are cases of causation, despite everyday talk and scientists' talk (see Mellor 1995: 52–53). But this would be an ad hoc response.

We now turn to the second premise, that the probability $P(E|C) = x$ is irreducible.[25] By 'irreducible' I mean probability that is an objective single-case chance,[26] which expresses an indeterministic fact about the world. If we take a probability p to be the application to an individual of a relative frequency R/S,[27] and suppose

$$R/S = p \qquad \text{where } p < 1$$

Then p is reducible iff there is a further factor A such that

$$R/S.A = q \qquad \text{where } q \neq p.$$

For example, the probability that I will get cancer given that I smoke is p. If there is a factor such as my vitamin A intake that, when conjoined with my smoking in the denominator, will give a lower probability than p, then p is reducible. Irreducible probabilities do not admit

25. This is a different sense of 'irreducible' from the one involved in the question of whether chance supervenes on particular matters of fact. See Dowe (unpublished-a) and Earman (1986).
26. Since we are considering singular causation, which is a relation between particulars, explicating the nature of the probabilities involved presents a serious difficulty. If causes supervene on laws, then the probabilities may be frequencies. But if we want a singularist account, the explication must allow for genuine single-case chance, which is a problem for frequency theories; but see Salmon (1988a). It must allow for objective chance, which is a problem for logical and subjectivist theories, which count probability as epistemic; but see Lewis (1986: chap. 19). On the other hand, there are difficulties in counting probability as a propensity (Dowe unpublished-a; Humphreys 1985). Of the different authors discussed in this chapter each has a different approach to probability. Since I will not be defending a probabilistic theory of causation, I offer no account of probability in the present work.
27. R is the 'attribute class,' and S is the 'reference class.' I use frequencies here just for the purpose of explication. There are well-known difficulties with the applicability of frequencies to the single case.

an ignorance interpretation – it is not possible by further investigation to discover further facts that will provide a better estimate of the probability.

Probabilities in quantum mechanics are generally thought to be irreducible in this sense. There is nothing about the Pb^{210} atom that can tell you any more about when it will decay, by what route it will decay (there are two possible routes for this decay, and the atom can also decay by β-decay to Bi^{210}), or whether for certain it will ever decay. And nothing about the environment of that atom can enable you to make those predictions, either. In fact, according to quantum theory there *is* no further fact that determines the result. We have here an irreducible probability.

One way out for the determinist,[28] and ergo, for the strict regularity theorist, is to deny the reasoning that adduces indeterminism from quantum physics. The alternative interpretative path is to treat the probabilities as reflecting ignorance, and to look for underlying deterministic regularities. However, such "hidden variable" theories came upon hard times with experimental proof related to Bell's Theorem (Bell 1964), which provided evidence that local hidden variable theories give wrong predictions,[29] and hence that the probabilities are irreducible.

However, the conclusion is not forced upon the determinist. There are other assumptions besides determinism involved in deriving Bell inequalities. It may be that an assumption that allows determinism is the one to go (see Earman 1986: sec. 10: 226–231). But even if that were so, the entire episode has nevertheless alerted us to the fact that causation should not be conceptually linked to determinism. Today, even if it turns out that we have no proof of indeterminism, we remain in the position of having learned that the concepts of determinism and causation are not conceptually linked.[30]

28. 'Determinist' here refers to one who denies that there can be any such thing as indeterministic causality.
29. This is widely held to prove that quantum mechanics is an indeterministic theory by proving the second premise in the argument: the state function in general only permits a prediction of the probabilities of the results of future measurements, and the state function is a complete description of the present state, therefore the present state does not completely determine future states.
30. There are various other motivations for treating causality probabilistically, besides indeterminism. Suppes shows that everyday causal language is usually probabilistic (1970: 7–8; see also Anscombe 1971), although a contrary view can be found in Mackie (1974: 49–50), and Salmon also concludes that determinism

If we accept both premises, then it seems the conclusion is forced on us. Causes need not be sufficient conditions for their effects. There are those who will deny this conclusion, for example, tobacco companies[31] and some philosophers (for example, Papineau 1989: 317–320; Rosenberg 1992; Russell 1948: 454), but it seems to be to be just reactionary thinking on both their parts. If philosophers still wish to insist that 'cause' means 'sufficient cause,' then we shall have to invent another word to denote the thing I am calling a cause.

II.3 THE COUNTERFACTUAL ACCOUNT

A different approach to causation – the so-called counterfactual theory – appears to overcome the above objection. This account was offered by Hume in the *Enquiry* as "other words" for his regularity theory: "Or in other words, *where, if the first object had not been, the second never had existed*" (*Enquiry*: 76, Hume's italics).

In the version of David Lewis (1986), c causes e just if c and e occur, and had c not occurred, e would not have occurred. The counterfactual "had c not occurred, e would not have occurred" is true if a possible world where c and e do not occur is more similar to the actual world than any possible world where c does not occur but e does. This counterfactual expresses a dependence of e on c that Lewis calls "counterfactual dependence."

This appears to solve the problem of indeterministic causation given in the previous section. There, we had the case where a lead atom decays, producing a mercury atom Hg, yet since not all lead atoms in similar circumstances decay to Hg, the regularity theory says, implausibly, that this is not causation. On the counterfactual account, however, it is a case of causation since had the lead atom not decayed, the Hg atom would not have existed.

But this appearance is misleading, for other kinds of indeterminism cause parallel problems for the counterfactual account. First, note that both the regularity and counterfactual theories are, in their own ways, deterministic. The essential difference is that the regularity theory takes

---

is an unnecessary burden on everyday usage. These arguments need not be adduced here, since we are offering a critique of Hume's *empirical* theory. Salmon (1984: 84–190) also gives examples from science to show that the same point applies there.

31. Who argue that smoking doesn't cause cancer because only very few smokers get cancer.

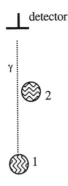

Figure 2.1. An indeterministic redundant cause.

causes to be sufficient for their effects, whereas the counterfactual theory takes causes to be necessary for their effects (although Hume's empirical analysis understands 'sufficient' in actualist terms, whereas Lewis takes 'necessary' in a modal sense as just described). This suggests that if the regularity theory has problems when the probability of the effect conditional on the cause is less than one, then the counterfactual account will have problems when the probability of the cause conditional on the effect is less than one, or in other words, when the probability of the effect without the cause is greater than zero.

For example, suppose a box contains two atoms, each with a chance of producing a gamma ray that might be detected by a detector. Suppose atom 1 produces a gamma ray at t_1, event c; that atom 2 does not; and that the detector detects a gamma ray at t_2, event e. Then it seems that c causes e, but it is not true that had c not occurred, e would not have occurred, because had atom 1 not emitted a gamma ray, atom 2 might have[32].

More specifically, a world W_1 where atom 1 does not emit, but atom 2 does, and a gamma ray is detected, seems to be just as similar to the actual world as a world W_2 where atom one does not emit, nor does atom 2, and no gamma ray is detected. Lewis has an account of what makes a world closer than another: a more similar world minimises widespread, diverse miracles, maximises regions of perfect match and minimises small local miracles, in that order of priority. However, our two worlds are equally similar to the actual world on these criteria,

---

32. There are alternative readings of might conditionals, but these are not relevant here (Barker 1998).

since neither has any miracles, and each has the same region of perfect match: the world up to the time of the cause. Lewis mentions another criterion, approximate similarity of particular facts, but doesn't endorse it. However, it is not clear that world W_2 is closer to the actual world than world W_1 on the criterion of approximate similarity of particulars. In W_2 there is no detection, in W_1 atom 1 emits a gamma ray. In any case, we could change the example so that the detection of a gamma ray would lead to a nuclear explosion, in which case W_1 is clearly closer to the actual world than W_2 on the score of approximate similarity of particulars.

Lewis has an apparent answer to this problem. He says that causation is either counterfactual dependence or the ancestral relation of counterfactual dependence – that is, where there is a chain of events linking c and e such that e counterfactually depends on the event before it, that event counterfactually depends on the event before it, and so on back to c. This at first appears to solve the case of the lurking back-up cause, because each stage of the gamma ray's transmission counterfactually depends on the previous stage, and the detection counterfactually depends on the final stage of the gamma ray transmission.

But this approach will always face a problem with the final step in the chain. The problematic case will be where the lurking back-up cause occurs at the very last moment. Then the final counterfactual dependence in the chain will fail to hold. For example, suppose atom 2 is immediately adjacent to the detector. Then a possible world where atom 1 does not emit, and where atom 2 does emit just after the actual time of the detection, and which does contain a detection (e) is at least as close to the actual world as any not-C world where there is no detection (late indeterministic redundant causation).

Deterministic cases are no problem, since given that the second atom did not actually decay in the circumstances, it would take a (small) miracle for it to do so, and hence such a world would be further from actuality than a world where neither atom decays, and there is no detection. It is only in indeterministic cases that such difficulties arise. (In section 2.5 we will consider Lewis's probabilistic theory.)

II.4 FREQUENT CONJUNCTION

One may be tempted to think that there is a simple way to modify Hume's theory (HEA) to avoid the above objection. In his influential book *A Probabilistic Theory of Causality* (1970), Patrick Suppes laments Hume's commitment to determinism: "The omission of probability considerations is perhaps the single greatest weakness in Hume's famous analysis of causality" (1970: 9). But elsewhere he says that "it is easy to modify Hume's famous analysis of causality in order to obtain a probabilistic characterisation of causes" (1984: 38). At one point Suppes proposes simply to replace constant conjunction by frequent conjunction, and interprets his own theory as providing precisely this generalisation:

Roughly speaking, the modification of Hume's analysis I propose is to say that one event is the cause of another if the appearance of the first is followed with a high probability by the appearance of the second. (1970: 10)

However, this is not what Suppes's actual theory gives us. Suppes's theory makes no use of the notion of frequent conjunction (that the conditional probability of the effect on its cause is high). As we shall see in section 2.5, his theory uses instead the notion that a cause raises the probability of its effect. This is not a frequent conjunction, for an effect may follow its cause with a low frequency. This surprising slip by Suppes is probably due to his eagerness to demonstrate his debt to Hume; and the fact that the debt is slightly more intricate than this does not lessen its significance.

While it is not Suppes's position, this *is* the modification that (on one reading)[33] Bertrand Russell proposed in his paper "On the Notion of 'Cause'" (1913), where Russell argues that the causal relation is "a frequently observed sequence" (1913: 2, 13), not a constant conjunction:

If stones have hitherto been found to break windows, it is probable they will continue to do so. ... We may then say, in the case of any such frequently observed sequence, that the earlier event is the *cause* and the later event the *effect*. Several considerations, however, make such special sequences very different from the traditional relation of cause and effect. In the first place, the sequence, in any hitherto unobserved instance, is no more than probable, whereas the relation of cause and effect was supposed to be necessary. I do not mean by this merely that we are not sure of having discovered a true case of cause and effect; I mean that, even when we have a case of cause and effect in our present sense, all that is meant is that, on grounds of observation, it is probable that when one occurs the other will also occur. Thus in our present sense, A may be the cause of B even if there are cases where B does not follow A. (1913: 12–13)

33. But see Anscombe (1971: 4, 6) and Dowe (1997b).

However, the frequent conjunction theory suffers the same fate as deterministic versions. In our case where Pb^{210} decays, the probability is very, very small – not anything like 1/2, let alone something closer to one. So, again, we have a case where a causal connection occurs without instantiating a frequently occurring sequence. The cause and effect instantiate a very rare sequence. In the vast majority of cases where one has hold of a lead-210 atom, one will hold it for a minute without its decaying. (Its half-life is about twenty-two years, so you'll have to hold onto it for most of your life to have any decent chance of its decaying.) In fact, in many examples of indeterministic causation the probability of the effect given the cause is actually low. Salmon has marshalled many examples, in the context of his discussion of explanation, that illustrate this point. (1970; 1971; 1984: 84–190). For example, we say that smoking causes cancer, although the chance is small that any given smoker will develop cancer.

It took philosophers in the twentieth century a long time to learn the lesson. In fact, one reason why causality fell into philosophical disrepute in the second quarter of the century was the widespread tendency to link causality and determinism. But even a casual survey of modern physics will show that, although indeterminism is widely accepted, the notion of causality remains. (See, for example, Suppes 1970: 6.) It's appropriate, then, that we turn now to the probabilistic account of causation.

II.5 PROBABILISTIC THEORIES

Patrick Suppes's work *A Probabilistic Theory of Causality* (1970) is a classic in the field. Suppes follows Hume in treating causality as a relation between events, except that the relation is a statistical one. To be more specific, the relation is one of positive statistical relevance (PSR). Suppes was not the first to attempt such a treatment – Reichenbach (1991) and Good (1961; 1962) both constructed theories based on positive statistical relevance[34] – but we will restrict our discussion here to Suppes's work, because his account has been more influential.

Suppes analyses causality in terms of probability theory in order to reflect the "wide use of probabilistic concepts in ordinary talk" (1970: 11) and because of the need "to include an appropriate concept of

34. A survey of these theories can be found in Salmon (1980).

uncertainty at the most fundamental level of theoretical and methodological analysis" (1984: 99). Suppes starts with the notion of an event as found in probability theory: an event is a set of possible outcomes, or more technically, a subset of the sample space. On this account 'events' are not always positive physical occurrences, but they are always related to a set of experimental data. The central intuitions that Suppes wishes to capture are: that a cause raises the probability of its effect, and that there is no third event that, when accounted for, nullifies (screens off) this positive statistical relation. To express this formally, Suppes defines prima facie causes and spurious causes in terms of probability relations among events, and then defines a genuine cause as a nonspurious prima facie cause:[35]

Definition 1. The event $B_{t'}$ is a *prima facie cause* of the event A_t if and only if

(i) $t' < t$
(ii) $P(B_{t'}) > 0$
(iii) $P(A_t|B_{t'}) > P(A_t)$

Definition 2. An event $B_{t'}$ is a *spurious cause* in sense one of A_t if and only if $B_{t'}$ is a prima facie cause of A_t, and there is an event $C_{t''}$ such that

(i) $t'' < t'$
(ii) $P(B_{t'}C_{t''}) > 0$
(iii) $P(A_t|B_{t'}C_{t''}) = P(A_t|C_{t''})$
(iv) $P(A_t|B_{t'}C_{t''}) \geq P(A_t|B_{t'})$

Definition 3. An event $B_{t'}$ is a *spurious cause* in sense two of A_t if and only if $B_{t'}$ is a prima facie cause of A_t and there is a partition $\Pi_{t''}$ such that

(i) $t'' < t'$

and for all elements $C_{t''}$ of $\Pi_{t''}$

(i) $P(B_{t'}C_{t''}) > 0$
(ii) $P(A_t|B_{t'}C_{t''}) = P(A_t|C_{t''})$

35. The following definitions are from Suppes (1970: 12–15).

Table 2.1. *Incidence of lung cancer*

	Vitamin A intake	
Age	Low	High
45–54	3 (690)	1 (1,807)
55–64	5 (1,122)	3 (2,365)
65–74	6 (830)	1 (1,464)
Urban	11 (996)	1 (1,915)
Rural	3 (1,672)	4 (3,721)
Total	2,642	5,636

Note: Number of men sampled in parentheses.

where a 'partition' is a collection of incompatible and exhaustive events.

The major reason that Suppes gives for including definition 3 as well as definition 2 is that the latter yields some undesirable results when one considers some of the more complex 'events' that arise on the set-theoretical approach (1970: 22–23). To illustrate these definitions, we will now consider a case (Bjelke 1975) mentioned by Suppes (1984: 49), which examined the effects of vitamin A intake on lung cancer among Norwegian men. Part of the experiment was to compare the incidence of histologically confirmed carcinomas other than adenocarcinoma (which I will henceforth refer to inaccurately as 'lung cancer') for men of various ages, places of residence and vitamin A intake. The relevant experimental data are shown in Table 2.1, showing the incidences of lung cancer, with the number of men sampled given in parentheses.

Let us define the events:

A – those who developed lung cancer
B – those with low vitamin A intake
C – those with high vitamin A intake
D – urban dwellers
E – rural dwellers
F – age 45–54
G – age 55–64
H – age 65–74

From these data we can easily establish that B (low vitamin A intake) is a prima facie cause of A (lung cancer), because B occurred before A, and P(B) > 0, and

$$P(A|B) = 0.0053 > P(A) = 0.0023.$$

However, B is not a spurious cause (sense one) of A because none of the available events, D, E, F, G or H, satisfy the conditions of definition 2. B is not a spurious cause (sense two) either, since neither of the two available partitions, $\Pi = \{D, E\}$ or $\Pi = \{F, G, H\}$, fulfils the conditions of definition 3. Therefore, according to the results of this experiment low vitamin intake (B) is a genuine cause of lung cancer (A), because B raises the probability of A, and there is no third factor that screens off this relevance.

Thus the PSR theory is built around the intuition that a cause raises the probability of its effect. (The further qualifications will not concern us here.)[36] We now turn to a line of criticism which I consider to be ultimately fatal to the probabilistic theory.

II.6 CHANCE-LOWERING CAUSALITY

One persistent argument against the PSR theory concerns counterexamples where a particular causal chain contains elements that do not stand in the PSR relation. Such counterexamples, which were the major point of dispute between Salmon (1984: chap. 7) and Suppes (1984: chap. 3), are variations on an example due to Rosen (1978). One such variation is the case where a golf ball is rolling towards the cup, but is kicked by a squirrel, and then after a series of unlikely collisions with nearby trees, ends up rolling into the cup (Eells and Sober 1983). This is a case, it is argued, where a singular cause lowers the probability of its effect, in other words, a counterexample to the claim that the PSR theory provides a necessary condition for singular causation.

I offer the following example (see Figure 2.2), which is an idealisation of a real physical nuclear decay scheme (Dowe 1993b), which is in turn a variant of a case presented by Salmon (1984: 200–201). This case, according to our best physical theories, involves genuine indeterminism.

36. Other important contributions to the probabilistic theory include Cartwright (1979) and Eells (1991). However, Cartwright's theory applies only to general causation, whereas our concern (in particular in relation to the following objection) lies with singular causation. Eells's account will be considered in Chapter 7.

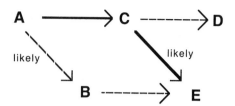

Figure 2.2. Chance-lowering decay scheme.

An unstable atom a may decay by various paths as shown in Figure 2.2, which depicts the complete range of physical possibilities for this atom. The event A denotes the existence of an unstable atom a, while B denotes the production of atom b, and similarly for C, D and E. (Note that lowercase denotes the atom, uppercase denotes the event.) We assume that each unstable atom has a very short half-life relative to the time frame under consideration. Note also that these events are not time-indexed.

The probabilities are as follows: $P(C) = 1/4$; $P(E|C) = 3/4$, and there are no other relevant factors. Take a particular instance where the decay process moves $A \to C \to E$. We would be inclined to say (correctly, I maintain) that the production of C is the cause of the production of E, in this instance. However, C and E do not stand in a relation of PSR, for $P(E|C) = 3/4 < P(E) = 15/16$.[37] So the cause lowers the probability of the effect. Thus we have a counterexample to the claim that the PSR theory provides a necessary condition for singular causation.

Furthermore, it is clear that we would regard the production of C as being of positive causal relevance, serving (in this case) to bring about the effect; and not as being of negative causal relevance, that is, serving to prevent the effect.[38]

In the *cascading* version of the decay case, we further suppose that time is discrete, and that the particular case involves cascading, that is, the two decay steps C and E occur at subsequent instances.

One variation of the PSR theory is the probabilistic dependence theory of David Lewis (1986: 175–178), which replaces the usual PSR relation with counterfactual conditionals about chance. These have the

37. Since $P(C|A) = 1/4$ and the schema is complete, $P(B|A) = 3/4$. Also, $P(E|B) = 1$. So $P(E) = P(E|C).P(C|A) + P(E|B).P(B|A) = 15/16$.
38. For the distinction between positive and negative causal relevance, see Humphreys (1989) and Eells (1991), and Chapter 7, where Eells's theory of probabilistic causality is discussed.

form "if event C were to occur, the chance of event E would be much greater than if C were not to occur." Lewis calls this relation 'probabilistic dependence.' This solves the case of indeterministic redundant causes, because it doesn't matter whether not-c worlds where e does not occur are closer than not-c worlds where e does occur because the other atom decays. The reason is that the chance of e at the time of c has the same value in both those worlds, and that value is less than the chance of e when c does occur.

However, while Lewis's theory has a number of advantages over the PSR theory, it too fails to account for the decay case of Figure 2.1. For if C were *not* to occur, that is, if atom c were not produced, then E would be more likely than if C *were* produced, because C's not occurring leaves a state of affairs such that A has not decayed, but will soon, ensuring that B occurs and subsequently, with probability 1, that E occurs.

To see this clearly, it must be appreciated that the decay schema is a schema, and not an 'evolution in time' representation. C is a causal interaction of shape Y, involving atom a instantaneously decaying into atom c (together with a β particle), while B is the decay of a into b (together with an α particle). A and C are *not* two interactions linked by a causal process depicted by the arrow. Nor are these events time-indexed. Thus, "if C were not to occur" means "if a were not to decay to c," so if C were not to occur then A would still obtain.[39] So Lewis's account fails to account for chance-lowering causes.[40]

Salmon has outlined a number of methods used to solve this type of problem (1984: chap. 7). The first method he calls *more precise specification of events*. The idea here is that a more precise specification of the cause, C, will yield the correct probability relations. In the case of the squirrel's kick, you might take account of the exact direction of impact of the kick. This is related to what Lewis (1986: 196) calls 'fragile' events. 'That the squirrel kicked the ball, imparting to it a certain momentum' is a more fragile event than 'that the squirrel kicked the ball.' Taking 'extreme standards of fragility' is demanding very closely specified events. However, there are no details about the production of atom c which will affect the probability relation.

39. Thus to say "if C were not to occur, B would occur" is not to violate Lewis's non-backtracking requirement (Lewis 1986: 35).
40. In Chapter 7 we will consider some of the modifications to this theory offered by Lewis.

There is no further fact about the atom which will enable us to know the outcome with more precision, if the standard quantum mechanical interpretation is correct. Thus there is no way to specify C more closely.

A more promising way, discussed by Menzies (1989a: 649), is to more closely specify the effect, E. Take E′ to be the production of atom e together with[41] an α particle, and E″ to be the production of atom e together with a β particle. (In Figure 2.1 the horizontal lines represent decay by β emission, and the angled line represents decay by α emission.) Then $P(E′|C) = 3/4$, but $P(E′) = 3/4 \times 1/4 = 3/16$, which is less than $P(E′|C)$. Hence C and E′ stand in the right PSR relation. So, according to the probability relations, B causes E″ and C causes E′; but we cannot say B causes E or C doesn't cause E, because this is description at the wrong level. The same result follows from Lewis's theory, since the chance of E′ if C were not to occur is zero.

However, there are some serious problems with this approach. First, in insisting on fragile events (closely specified events) we are moving away from ordinary causal talk. Normally we would suppose that if C caused E′, then C caused E (where E′ is a closer specification of the event E). If smoking caused my lung cancer together with some other side effect, then we happily say that smoking caused my lung cancer, *simpliciter*. (See Lewis 1986: 198; Menzies 1989a: 649). Or if my cancer caused me to die in pain, then my cancer caused me to die. So at least sometimes in ordinary causal talk there is an entailment concerning the effect-event: where the fine-grained event is caused, the coarse-grained event is caused. This relation is not mirrored by an account that insists on fine-grained effect-events. Thus it follows from this approach that much of our causal talk is literally false, since it does not describe events to a sufficient extent. Given that we are seeking an empirical analysis, that we move away from ordinary causal talk is not in itself such a bad thing, provided we have sound physical reasons. But there is no clear physical picture here to appeal to.

Second, there is no principled basis for deciding how fragile an event need be. How closely do events need to be specified? There seems to be no nonarbitrary answer to this (the so-called aspects problem – see Davidson 1969; Kim 1969). This is especially relevant here, since it

---

41. 'Together with' means in part that the e-atom and the α particle appear together at the same time. This restriction or an equivalent condition is necessary to distinguish the α particle from the one that is produced in the a to b decay.

means that the rescue strategy is entirely ad hoc (Menzies 1989a: 649–650).

Third, there are problems for counterfactuals, as Lewis has shown (1986: 198). In Lewis's example, a firing squad of eight soldiers kills someone. Under requirements of extreme fragility of effects, had one more soldier also fired, the resulting death would have been a different event from the actual death. Suppose the gentle soldier did not shoot. Then his failure to shoot makes the death a different death from the one that would have occurred had he fired. This makes the soldier's omission a cause of the death.

Salmon's second method is the method of *interpolated causal links*. Here one finds a further event, between the cause and the effect, which when conditionalised on will provide the right probability relations. For example, we should account for the golf ball's motion just after the squirrel's kick. Then, given that event, it is more likely that the ball will sink if it hits a tree than if it doesn't. In the decay case it is not so easy to find the right intermediate event. We could include the event A', the decay of A by β decay. But then $P(E|C.A') = P(E|A')$, which means that C is causally irrelevant. (One problem here may be that A' and C are not distinct events.) Generally, in the decay case we find that the right sort of events are not available, as Salmon has argued.[42] In particular, in the cascading case there simply is no such event.

Salmon's third method is the method of *successive reconditionalisation*. (See also Good 1961; 1962.) In this method we consider the probability of each event in a chain, conditionalised on the previous event. Suppose, in the squirrel case, we have the series of events: A – the ball is rolling towards the cup; B – the squirrel kicks the ball; C – the ball hits a tree; D – the ball rolls into the cup. Then we have the inequalities, $P(B|A) > P(B|\sim A)$; $P(C|B) > P(C|\sim B)$; $P(D|C) > P(D|\sim C)$, where '$\sim A$' denotes 'not A.' A can be said to be a cause of D because it is an indirect cause (see Salmon 1984: 200). However, in the cascading decay case this method will not work because there is no intermediate event between the cause and the effect. In fact, the importance of the decay case is precisely that it cannot be solved by any of these methods, whereas cases such as the squirrel's kick can.

So it seems that this decay case is a genuine counterexample to the PSR and counterfactual probabilistic dependence theories. It

---

42. This and the next strategy will be considered in more detail in Chapter 7.

follows that they do not provide a necessary condition for singular causation.

One popular reply, suggested by Suppes (1984: 67) and defended by Papineau (1989; 1986) and Mellor (1995: 67–68), is to claim that in cases where a cause lowers the probability of its effect we say that the second event occurred 'despite' the first; and so we should follow our intuitions and admit that it is not a cause. 'Despite,' like 'prevent' and 'inhibit,' signifies a negative cause.[43] Call this the *despite defence*. If $P(B|A) <$ $P(B)$ and A and B occur, then B occurs *despite* A and can be said to be a negative cause, *inhibiting* B. Thus the golf ball sank *despite* the squirrel's kick, and atom e was produced *despite* the decay of atom c. Others claim that our intuitions are unclear at this point and that the issue needs to be solved elsewhere (for example, Mellor 1988), that is, it's spoils for the victor.

I have three objections to the despite defence. First, it is not consistent with the way physicists speak: "the production process here involves C producing E" (see Enge 1966: 227–228), where 'produce' is clearly a near-synonym for 'cause,' that is, a positive cause. Clearly, scientists think of the process in terms of positive causation.

Second, 'despite' does not unambiguously denote a negative cause. We are concerned with the sentence, "B occurs despite the occurrence of A." Fundamentally, 'despite' indicates surprise: the joint occurrence was not expected, or is unlikely (see Menzies 1989a: 648). This can be read as $P(B|A) < P(B)$, a relation of negative statistical relevance. Therefore we can say that at the general or class level A is a negative cause of B. But 'despite' may also indicate a negative cause: an inhibitor or preventer. For at the singular level A may be either a negative or positive cause. That is, 'despite' can indicate either a negative cause that served to *inhibit* the occurrence of B, or an actual positive cause that was unexpected. Consider the following two sentences:

(i) I caught a cold despite taking vitamin C tablets.
(ii) Lara was clean bowled despite the fact that the bowler was Taylor.

In (i) 'despite' indicates a negative cause, since vitamin C, even in this instance, served to (unsuccessfully) inhibit the cold. In (ii) 'despite' indicates an unexpected positive cause, because Taylor was the actual

---

43. A discussion of this distinction may be found in Humphreys (1981). See also Eells (1991), and my discussion of Eells in Chapter 7.

cause responsible for Lara's downfall (surprisingly enough, since Taylor is not in the team for his bowling). At the general level, both would be governed by a relation of negative statistical relevance. Therefore we must conclude that the 'despite' argument is flawed: 'despite' does not unambiguously denote a negative cause.

At least, more work is required. The mere fact that we would say "B occurred despite A" does not indicate that we think that A was not a cause of B. We still need to ask whether in the *particular* case A causes B or not. Further consideration from the context is always necessary. Take the decay case. Eells (1988: 130) and Menzies (1989a: 660) claim that this type of case is a 'despite' construction and so is a negative cause. I agree that we may say E occurs despite C, but in this instance I claim C causes E, like (ii), and 'despite' indicates surprise – the particular case of C causes E even though in general it makes E less likely.

Third, and most significantly, decay chains which bear the PSR relation and chains that do not (such as the case described earlier) can have identical mechanisms, and therefore in such cases there is no ground for distinguishing 'because' and 'despite.' For example, take the particular instance $A \rightarrow C \rightarrow E$. This is not a causal chain, according to the despite defence, since C does not stand in the appropriate relation to E; E occurs despite C. But take the particular chain $A \rightarrow B \rightarrow E$. This is a causal chain, since B and E stand in the right relation. My argument is that the difference between these two chains does not warrant the distinction between causal and not causal. One difference is that $A \rightarrow C \rightarrow E$ proceeds by β decay, followed by α decay, while $A \rightarrow B \rightarrow C$ proceeds by α decay, followed by β decay. But since this involves identical processes, just in a different order, the distinction cannot be grounded in this difference. In other words, the two processes are the same, but in different order.[44] Thus the despite defence forces us into a very implausible judgments about such mechanisms.

Mellor has defended the despite defence by drawing a distinction between particular and factual causes (1995: 67–68). The particular event Don's fall is not the same thing as the fact that Don fell, although both may be causes. Mellor considers a standard chance-lowering example. Sue pulls her golf drive (p), but the ball hits a tree and bounces into the hole (h). "But the pulled drive causes her to hole out

44. A more detailed version of this argument, based on a real case, is given in Dowe (1993b).

in one, even though she would have had a greater chance of doing so had she not pulled her drive" (1995: 68).

But, says Mellor, we must distinguish the particular event Sue's pulled drive (p) from the fact that she pulls the ball (P). In the quoted intuition just given, the causal claim concerns causation between particulars, a claim that is true, as is the causal claim that Sue's driving the ball (d) causes h, not surprising since d and p are the same particular. But the chance relation in the quoted sentence concerns facts, as do on Mellor's view all chances. The fact P – that Sue pulled her drive – is not the same fact as the fact D – that she drives her ball. P lowers the chance that she holes in one (H), and is not a cause; whereas D raises the chance of, and causes, H.

This blocks putative chance-lowering causation between particulars. But it does not block the intuition we have about facts. In this case Sue holed in one (H) because she pulled her shot (P), yet P lowers the chance of H.

So it seems that the despite defence does not work, and there is a genuine counterexample to the claim that the probability relation approach provides a necessary condition for singular causation. Like the constant conjunction and frequent conjunction theories, the probabilistic theory fails to account for certain cases of genuine indeterministic causation.

II.7 SUMMARY

I have argued that the deterministic, counterfactual and probabilistic accounts fail to provide an account of singular causation as it is in the world. In particular, probabilistic theories, taken as aiming to provide an empirical account of singular causation, fall to an important counterexample from subatomic physics. Attempts to avoid this counterexample have been carefully examined and refuted.[45]

We therefore turn in the next chapter to a non-Humean approach – one that takes causation to be about the causal connection understood in terms of physical quantities such as energy. I shall argue that this sort of theory easily accounts for the indeterministic and chance-lowering examples.

45. Although we return to this counterexample in the second half of Chapter 7, when we look at attempts to combine the insights of Chapter 5 with the probabilistic account.

3

Transference Theories of Causation

In this chapter I examine non-Humean 'transference' accounts of causation due to Aronson and Fair. I show that this kind of account fits well with the cases of indeterministic causality and chance-lowering causality discussed in the previous chapter. I also propose a way to extend those accounts into a full conceptual analysis, thereby developing the idea of transference as a meaning analysis.

However, there are a number of difficulties with the transference account. These concern problems of the identity over time of the transferred quantities, and the direction of causation. Further, I argue that there is a kind of causation, immanent causation, or causation as persistence, which is neglected by the transference accounts. I spell these out these concerns in section 3.4.

III.1 REVIEW OF TRANSFERENCE THEORIES

The key idea of transference theories of causation is that a cause transfers something like energy to its effect, which makes causation a singular matter. Although this cannot be what we mean when we use the word cause, it is a plausible candidate for an empirical analysis, and one which has been widely held in previous centuries. In this chapter we consider the two most influential proponents of this approach, Ron Aronson and David Fair.

III.1.1 Aronson's Transference Theory

Aronson's theory was first offered in two papers in 1971, in deliberate contrast to manipulability accounts (1971b) and regularity accounts (1971a). The theory was presented in three propositions:

(1) In '*A* causes *B*,' '*B*' designates a change in an object, a change which is an *unnatural* one.

(2) In '*A* causes *B*,' at the time *B* occurs, the object that causes *B* is in contact with the object that undergoes the change.

(3) Prior to the time of the occurrence of *B*, the body that makes contact with the effect object possesses a *quantity* (e.g., velocity, momentum, kinetic energy, heat, etc.) which is transferred to the effect object (when contact is made) and manifested as *B*. (1971b: 422)

Proposition (1) refers to a distinction Aronson draws between natural and causal changes – causal changes are those that result from interactions with other bodies; natural changes are not causal, and come about according to the normal course of events, when things happen without outside interference. Thus internal changes, or developments, are not seen by Aronson as cases of causation. Proposition (2) is Hume's requirement that causation occurs only by contact, which rules out action at a distance. It also means that, strictly speaking, there is no indirect causation, where one thing causes another via some intermediate mechanism. All causation is direct causation.

Proposition (3) is the key notion in Aronson's theory. It appeals to the idea of a quantity, which is possessed by objects, which may be possessed by different objects in turn, but which is always possessed by some object. (Harré suggests that it could be understood in terms of 'substance' [1985: 485].) This quantity is transferred from cause to effect. There is, says Aronson, a 'numerical identity' of the quantity involved in transference. However, Aronson does not give a definitive description of what quantities may play this role, but is content to give a list of some examples: velocity, momentum, kinetic energy, and heat.

The direction of transfer sets the direction of causation. This is not necessarily connected to the direction of time, however. Aronson explicitly denies that causation is to be taken as conceptually connected to time in that way. He considers the case of two connected gears, where one drives the other, a case of simultaneous causation. Energy flows from one to the other, from the cause to the effect, but one is not temporally prior to the other.

Aronson's major argument for his theory concerns what he calls the "grammatical or syntactical" features of 'cause' (1971b: 417). 'Cause,' he argues, is essentially a "dimension-word" for a range of transitive verbs, such as 'push,' 'hit,' and so forth (but not 'know' or 'see'). 'Cause' can always be replaced by one of these transitive verbs, and these transitive verbs can always be replaced by 'cause.' What all these transitive verbs have in common is that they describe a transfer of some quantity. So, mechanical causation is the core notion, and the word comes to be applied to mental phenomena by analogy to the mechanical case (that is, the manipulability theory has the relation back to front). Other accounts of causation cannot account for the place that transitive verbs play in the core instances.

The problem of the relation between ordinary concepts and energy and/or momentum transference is illustrated by an objection Earman levelled against Aronson. Earman (1976: 24) asks us to consider a possible world quite unlike ours, where our conservation laws do not hold. In that world a moving ball strikes a stationary ball, and following the collision the moving ball continues on with the same momentum as before the collision, while the stationary ball moves off with some velocity. No momentum has been transferred, since the original ball still has all of its momentum. Yet we would say, Earman urges, that the moving ball causes the stationary ball to move. Therefore "our intuitions are not founded on and justified by Aronson's transfer-of-quantities story" (1976: 24). This is a serious point for anyone who is offering an account of the way we use language, but Aronson is not open to such criticism, since there is no implication here that transference is simply what we mean. While it is not clear in Aronson's earlier papers (which Earman was discussing), Aronson's later response makes it quite clear that his account also is to be understood as an ontological account, not a conceptual account (1982: 294, 302).

III.1.2 Fair's Transference Theory

In 1979, David Fair offered an account of causation similar in many respects to that of Aronson, although the details of Fair's version are worked out more carefully (1979). Fair makes the claim that physics has discovered the true nature of causation: what causation really is, is a transfer of energy and/or momentum. This discovery is an empirical matter, the identity is contingent. Fair presents his account as just a program for a physicalist reduction of the everyday concept, and he doesn't claim to be able to offer a detailed account of the way energy transfer makes true the fact that, for example, John's anger caused him to hit Bill. A full account awaits, Fair says, a complete unified science (1979: 236).

Fair's program begins with the reduction of the causal relata found in ordinary language. Events, objects, facts, properties and so forth need to be redescribed in terms of the objects of physics. Fair introduces "A-objects" and "B-objects," which manifest the right physical quantities, namely energy and momentum, and where the A-objects underlie the events, facts, or objects identified as causes in everyday talk, while the B-objects underlie those identified as effects. The physical quantities, energy and momentum, underlie the properties that are identified as causes or effects in everyday causal talk.

The physically specifiable relation between the A-objects and the B-objects is the transfer of energy and/or momentum. Fair sees that the key is to be able to identify the same energy and/or momentum manifested in the effect as was manifested in the cause. This is achieved by specifying closed systems associated with the appropriate objects. A system is closed when no gross energy and/or momentum flows into or out of it. Energy and/or momentum transfer occurs when there is a flow of energy from the A-object to the B-object, which will be given by the time rate of change of energy and/or momentum across the spatial surface separating the A-object and the B-object.

The reduction thus stands as:

A causes B iff there are physical redescriptions of A and B as some manifestation of energy or momentum or [as referring to] objects manifesting these, that is transferred, at least in part, from the A-objects to the B-objects. (1979: 236)

Fair's major argument for the transference theory (whether it be Fair's version or Aronson's) is that it explains the fact that we all manage to agree about most cases of causation. The manipulability account, for example, cannot explain that fact. But if causation really is a transfer of energy and/or momentum, then it is an objective feature of the world, and even if we don't realise that causation really is energy and/or momentum transfer, if we have reliable ways to identify causation then it will be no surprise that there is widespread agreement. We do have reliable ways to identify causation, Fair says, through an appeal to features such as contiguity and regularity. The transference theory explains why those are reliable indicators. So the transference theory explains why there is widespread agreement.

III.1.3 Comparison

There are several points of difference between the accounts of Aronson and Fair. First, Fair does not require a 'full identity' of the quantity. Since in real situations energy tends to be dissipated, there is never a full conservation of the relevant quantities. So Fair allows that there be a partial transfer of the quantity, holding that conservation of the quantity is not a conceptually necessary part of causation. Second, Fair is a lot clearer about what quantities are relevant. Aronson says "velocity, momentum, kinetic energy, heat, etc.," but this does not really tell us what sort of quantities we are talking about. Fair is a lot more explicit about what gets transferred: the relevant quantities are just energy and/or momentum.

I think Fair is right here. As Fair says, what about the quantity 'negenergy,' the energy lost in an interaction? If negenergy counts as a quantity that can be transferred, then effects become causes and causes become effects.

Or consider a light show where there are two coloured spots on a screen. Suppose the red spot is moving towards the stationary blue spot, and exactly as the red spot reaches the point where the blue spot is, the red spot stops and the blue spot moves off. Is velocity transferred from the red to the blue spot? Nothing in Aronson's theory tell us that it isn't, and if so then the red spot caused the blue spot to move, on Aronson's account. The moral: velocity cannot be the right kind of quantity.

Another point of difference concerns the asymmetry of causation. Aronson says the asymmetry of causation is explained by the asymmetry of transference: a quantity is transferred *from* the cause *to* the effect. In this Aronson deliberately avoids appeal to the time order of the events. Fair, on the other hand, defines causation in terms of temporal order: transference is defined in terms of the time derivative of the energy or momentum. So cause and effect are distinguished by virtue of their being earlier and later, respectively. We shall return to this issue.

III.2 TRANSFERENCE, INDETERMINISM AND CHANCE LOWERING

In the previous chapter we found that Hume's regularity theory, being a deterministic theory, does not allow for cases of genuine indeterministic causation, such as the case of the decay of Pb^{210} by α-decay to Hg^{206}. We also found that Suppes's probabilistic theory does not account for genuinely chance-lowering cases, such as the case depicted in Figure 2.1, where c decays by α-decay to e.

Transference theories have no difficulty with such cases. Taking the simple indeterministic case first (the decay of atom c by α-decay to e), we note that the energy of the lead atom is partly transferred to the mercury atom and partly carried off by the alpha particle. According to the transference theory, then, the existence of the lead atom is the cause both of the production of the mercury atom and of the existence and the energy of the alpha particle.

In the chance-lowering case, the same story applies. The energy of the c-atom is partly transferred to the e-atom and partly carried off by the alpha particle. It is not relevant that the probability of e being produced is lower given the existence of the c-atom. Energy is transferred, so it is a cause. Thus the transference theory has a great advantage over regularity and probabilistic accounts – it allows for the cases of genuine indeterministic causation and genuine chance-lowering causation.

III.3 TRANSFERENCE AS MEANING ANALYSIS

Aronson and Fair are very clear about the status of their proposed identity – it is an a posteriori discovery. It is not something that a competent user of the language could necessarily be expected to know about. Without question they are right. But perhaps the basic notion involved in the account can be used to extend the account to include an account of meaning. For while it is no part of our ordinary notion that causation involves transfer of energy or momentum, it's quite plausible that our ordinary notion could profitably be subsumed under the notion of giving, or transference *in general*. Then the discovery is that in our world the thing that is given or transferred is energy or momentum. This is not an approach that I will be advocating in this book. As we will see, I think there are serious objections to transference accounts. But it is of interest in the present context to see how the account could be extended.

The proposal is to treat causation as a theoretical term, irreducible, especially not reducible to anything like energy or momentum. This approach enables a meaning account to be given that avoids standard sorts of difficulties.

The proposal begins by noting that standard philosophical theories of causation, while containing important insights, never seem to capture the essence of folk meaning. For example, the regularity theory, if intended as meaning analysis, clearly fails, for folk identify causes while having hazy or no knowledge of the relevant regularities. Take, for example, Gasking's case of dropping a white pellet into a glass of

46

water, with an ensuing explosion. Folk will say that dropping the pellet into the water caused the explosion, without any knowledge of what usually happens with that particular substance (whatever it is). Similarly, the manipulability thesis, if taken as meaning analysis, fails to pick up adequately the idea that there is physical causation that occurs quite independently of agents and their intentions, sometimes quite beyond the reach of any human agents.

One influential approach to analysis that combines the conceptual and empirical analyses, yet avoids the sort of problem just mentioned in regard to the regularity and manipulability theses, is the Ramsey-Lewis approach to theoretical terms and the associated style of analysis. That approach treats the concept for analysis as an irreducible theoretical term. But irreducible theoretical terms may nevertheless be analysed along the lines of the Ramsey-Lewis approach. The idea is that we think of our ordinary concept as being the subject of a theory – the 'folk theory' – involving certain theoretical commitments about the nature of the concept. These commitments, called 'the platitudes,' may be conjoined into one long sentence called by Lewis "The Postulate" of the folk theory, which contains the folk intuitions about the concept. The Postulate can be expressed in a certain logical form, the 'modified Ramsey sentence,' which implies a unique realisation.[1] In effect this expresses an (a priori) functionalist conceptual analysis of the relevant discourse, in the sense that the concepts of the folk theory are defined functionally: such and such is the thing that fulfils such and such a function in the theory. This functional definition is then open to a posteriori identification: such and such an item in the world is the thing that has such and such a function in the theory.[2]

For example, the Armstrong-Lewis approach to the mind takes mental concepts to be defined as the states that play certain roles – states apt to bring about certain responses given certain stimuli. This 'causal theory' of the meaning of mental terms is then supplemented by the 'identity theory,' which identifies those states as being certain

---

1. Let T be the folk theory of the relevant discourse, where the relevant predicates are written in property-name style and replaced by a variable: $T(x_1, x_2, \ldots)$. The Ramsey sentence of T is $(\exists x_1)(\exists x_2) \ldots T(x_1, x_2, \ldots)$, and the modified Ramsey sentence of T is $(\exists x_1)(\exists x_2) \ldots (y_1)(y_2) \ldots \{T(y_1, y_2 \ldots) \text{ iff } x_1 = y_1 \,\&\, x_2 = y_2, \,\&\, \ldots\}$. See Lewis (1983: chap. 6).
2. This identification, if true, will be a necessary truth if construed as referring to the thing that plays that role in our world; and a contingent truth if it refers to whatever plays that role.

47

neural states in the brain – a contingent identity thesis. This adds up to the thesis that mental states are neural states.

This general approach to analysis has been applied to causation by Michael Tooley in his book *Causation: A Realist Approach* (1987: 251), and also by Peter Menzies in a paper entitled "Probabilistic Causation and the Pre-emption Problem" (1996).[3] Tooley and Menzies construe causation as an irreducible theoretical concept, but believe it can be analysed in the Ramsey-Lewis style.[4] On Tooley's analysis, causation is the theoretical relation that determines the direction of the logical transmission of probabilities. On Menzies's analysis, causation is the intrinsic relation that typically exists between two events just when the events are distinct and one event increases the chance of the other event (Menzies 1996: 24). This opens the way for a theoretical identification to be made – Menzies suggests as a plausible possibility that the relation between chance-raising events is in fact a relation of transference of energy or momentum. Then the identity is given that causation is energy transference.

Although there may be advantages to the Ramsey-Lewis approach, this particular version faces problems. For example, important though the chance-raising notion is, it seems quite implausible that this is what folk mean by their causal talk. It is most likely that folk don't understand the distinction between 'chance-raising' and 'probable.' They cannot reliably distinguish between the two types of cases. Even Patrick Suppes confuses the two in the introduction to his classic work on the subject! (1970: 10) Nor is it possible that folk have just *tacit* knowledge of chance-raising. This is shown by the failure of standard decision theory as a description of how people actually make decisions.[5] But folk can reliably identify cases of *causation*. Therefore, chance-raising cannot be an essential element of the folk theory of causation.[6]

What I am proposing is a different analysis, but also along the lines of the Ramsey-Lewis approach. The proposal is that central to The Pos-

---

3. Also see Jackson's application of the method to moral discourse (1992).
4. Of course, if the Lewis-style functionalist account of the mind is being generalised and applied to causation, then the 'functionalist' account of the role causal terms play in folk theory cannot invoke *causal* roles, for that would be circular. Presumably the roles performed will be inferential.
5. Although Ramsey, who offered the original version of standard decision theory in 1926, conceived of it as a description of how people actually make decisions.
6. In Chapter 7, section 7.2, a different argument is urged against Menzies's theory, one that draws on chance-lowering causality.

tulate is that an object, the cause, possesses something, which it transfers to another object, which subsequently possesses the same thing. Causation, then, is the relation that holds just when there is something that is possessed by one object (the cause), and transferred to another object (the effect). Then, as a contingent identity, it is proposed that the thing transferred in our world is energy or momentum. Therefore, in our world causation is energy or momentum transfer. The idea may be expressed as follows:

Meaning Analysis: Cause and effect are whatever are involved in giving.
Contingent Hypothesis: The thing transferred is energy or momentum.
A Posteriori Identity: Causation is energy or momentum transfer.

Is this a plausible meaning analysis? While I have no interest in offering a detailed defence of the idea, there are some considerations worth noting, which indicate that this is a plausible meaning analysis. In Aronson's original defence of the identification of causation with energy/momentum transfer, 'cause' is treated as a dimension word of a class of certain transitive verbs – such as 'hit,' 'push,' and so on. 'Giving' is one of the transitive verbs that Aronson discusses. At least in those cases the transference account is perfectly natural: "a gave b the measles," or "the volcano gave off smoke and small rocks," can be recast using the word 'cause': "a was the cause of b's getting measles," "the volcano caused smoke and small rocks to enter the atmosphere." Notice that 'giving' can involve agent *or* mechanical causation.

However, Aronson takes the mechanical cases to be the fundamental notion, with the word used in the case of agent causation "in analogy." This is the reverse of the path taken by the manipulability account of causation, which takes agent causation as fundamental. (See Gasking 1955; Menzies and Price 1993; von Wright 1971; 1973.) Both approaches seem questionable at the level of meaning analysis. On the one hand, we think of ourselves as causal agents with genuine power as initiators in a way fundamentally different from the way a rock can be a cause. It can't be a platitude of the folk theory that agents are just like rocks (even if it is true). Aronson's claim that mechanical cause is the fundamental notion must be utilised at a different stage in the analysis, as Aronson later came to appreciate (1982; 1985). On the other hand, manipulability theories, which take agent causation as fundamental, face the difficulty that much causation in the world occurs beyond the reach and power of human agents – billions of light years beyond, in many cases. When we say the cause of the carbon in the earth's environment was an explosion of such and such a star, so many billions of years ago, it cannot be that we are really thinking of the cause as a way for agents to bring about that sort of effect.[7]

However, 'giving' seems to involve the appropriate ambiguity between agent and physical causation. Giving can be intended, or accidental. Either way, it's still giving. That seems to be an advantage over both the energy/momentum transfer account (which is acknowledged to fail as a meaning analysis) and the manipulability account (which purports to be a meaning analysis).

There is more that can be said in defence of transference as a meaning analysis. In a 1988 paper, Mellor sets out what he sees as the "connotations of causation." In so doing, Mellor proposes, as the correct approach to settling disputes about concepts, to follow the dictum, "Don't ask for the use, ask for the point of the use" (1988: 230). Mellor's 'connotations of causation' are therefore the sort of things that, if his analysis is right, will figure as platitudes of the folk theory. They are not simply cases of usage, but universally recognised theoretical guidelines for the use of the concept. Let's test the current proposal against Mellor's connotations. He says, "Causation's main connotations are the following" (1988: 230; see also 1995; I shall take them in a different order than does Mellor):

(1) "*Means-end*: If an effect is an end, its causes are means to it." Mellor goes on to show how this notion may be considered as the fundamental one. I have just indicated why I think that is mistaken. But that is not to say that (1) is not a platitude with widespread relevance. But if it is, then the notion of giving perfectly explains why it is: having something and being in a position to give it to someone or something precisely *constitutes* means to an end, namely the end of that someone or something coming to have that thing.

(2) "*Evidential*: Causes and effects are evidence for each other." If a had something and was in a good position to give it to b, and then no longer had that thing – that in itself constitutes evidence that b came to have that thing. On the other hand, if b came to have something, and a had been in a good position to have given it to b, that constitutes evidence that a had, and gave up that thing.

7. For a counterargument, see Menzies and Price (1993).

(3) "*Explanatory*: Causes explain their effects." Similarly, if a had and gave up something, that explains why b came to have it. Of course, that b came to have it doesn't explain why a had it and gave it up.

(4) "*Temporal*: Causes precede their effects." This, I think, is questionable. For a start, some of the folk platitudes involve causes contemporary to their effects, such as Aronson's example of the engaged gears. They move at the same time, but one is the cause and the other the effect. Perhaps this is not the correct physical analysis – but that doesn't mean it's not among the folk platitudes. Further, I think most folk would regard causal connections such as backwards-in-time voodoo as envisaged by Dummett (1964) as probably false, but nevertheless coherent. So I take it that (4) is not one of causation's main connotations. However, it may well be part of the folk theory that (4) is true most of the time. As a fairly widespread fact about causation, rather than an essential part of its meaning, causes precede their effects. There is no reason why the notion of giving is incompatible with that fact.

What is a connotation of causation is its asymmetry. Causes produce their effects and not vice versa. It does not follow from "a caused b" that "b caused a" but rather that "b did not cause a." The relation is asymmetric, unlike a symmetric relation such as, say, 'is causally connected to.' So causation has a direction. But this direction is not conceptually connected to the direction of time, at the level of folk usage. Whether causes occur before, simultaneous to, or after their effects there is still a direction to the causal relation. So I propose that (4) be replaced by the following:

(4′) *Asymmetry*: Cause produce their effects and not vice versa. This is clearly reflected in the notion of giving or transference, which also has a directionality. Being the donor, and being the receiver, are not interchangeable. 'To transfer,' like 'to cause,' has a direction built into its meaning.

In summary, the notion of giving fits well with Mellor's connotations, appropriately refined, although a more detailed analysis would be required before conclusions could be drawn with confidence. However, I will not examine this proposal any further, because I think there are some telling objections to the whole transference approach. We now turn to these.

III.4.1 Persistence as Causation

There is a type of causation quite different from the type envisaged by the transference theorists, and one that the transference theory rules out of court. This is the case where an object persisting in time is thought of as a causal process. One persuasive example is when an object's inertia is the cause of its continuing motion. For example, consider a spaceship moving through space with constant rectilinear motion, not acted on by any force. We should say that the cause of its continuing motion is its own inertia, and indeed, that earlier states are the causes of the later states. But there is no transfer of energy or momentum from one object to another – in fact, there is nò causal interaction. That it fails to allow for such cases of causation is a shortcoming of the transference theory.

It seems that there are two types of causation, the type where there is a transfer, and the type where there is a persistence of a certain sort. These twin notions can be found in Spinoza, who distinguished between an *immanens* cause and a *transiens* cause (Bennett 1984: 113), and in Leibniz, who distinguished between intrasubstantial and intersubstantial causation. David Armstrong discusses a type of causation where temporal parts of an individual are connected by causal relations (1980), and uses that notion to construct an account of identity through time.

It should not be supposed that this objection is motivated by intuitions about the folk concept. Not so. In fact, it is a little awkward to say that an object is the cause of its being there later. No, in my mind the objection arises from a consideration of what science teaches us about how causation works, as we shall see in the subsequent argument.

Similar objections to the transference theory have been made before, although not in any detail. Beauchamp and Rosenberg have offered, as a counterexample to Aronson's theory, the case of a moving pendulum, describing it as a causal sequence involving a single object (1981: 208).[8] I agree that this is a causal process, but the transference theory is not threatened, because the pendulum moves under the influence of the gravitational field, so there is a transfer of energy/momen-

---

8. Since developing this argument I have found that Tooley makes the same point independently (1993). In any case, I suspect that I got the idea from D. M. Armstrong; see Armstrong (1997).

tum. (No doubt fields should be construed as objects for the purposes of this theory.)

Ehring offers the example of a chemical substance undergoing "internal change" (1986: 250). Aronson replies: "Pray tell, how can a chemical substance such as a gas undergo a change of internal state unless something is doing work on it or it is performing work on something else?" (1985: 250–251) This reply is a bit hasty, for there are processes that change without external influence – in fact, Ehring gives the example of nuclear decay. A better answer on Aronson's part might have been to point out that such chemical processes can be broken down to their parts, for example, molecules, at which level we discover a whole lot of causal interactions occurring, each involving energy/ momentum transfer.

But neither of these defences works in the case of the spaceship moving by its own inertia. It is not moving under the action of an external field, and breaking it down to its parts will reveal that the causal story cannot be retold entirely in terms of micro-transference.

Another strategy available to the transference theorist is to deny that persistence is a type of causation at all. This appears to be the attitude of Aronson and Fair. But in addition to the inherent plausibility of the spaceship case, there are some other considerations that add further weight.

The spaceship moves according to Newton's First Law, which states that a body will continue in motion unless acted on by a force. But consider Aristotle's physics. According to Aristotle, a body requires a force for it to continue in motion. According to Aristotle, then, the spaceship comes to a halt, or if it does continue to move, then there must be some external force acting. So in the case where the spaceship moves with constant motion, Newton and Aristotle disagree about what is the cause of that motion. Newton says the cause is the body's inertia,[9] Aristotle says the cause is some unknown field. Both offer causal explanations of the motion. To say, as one would if one were to deny that the body's inertia is a cause of its own motion, that Newton provided no causal explanation is to take special pleading too far, in my view.[10]

9. In Chapter 5 we will consider the significance of the fact that in relativity, velocity is frame-dependent.
10. Special pleading entailing that according to Newtonian mechanics nothing is responsible for the body's motion. A person so steeped in ordinary ways of thinking about causation that she cannot accept this argument might like to ask herself

One common response to this argument is to say that the point of Newton's analysis is to show that uniform motion is not caused, because there is no change occurring. However, this does not take into account the importance of the concept 'inertia.' As one dictionary of physics puts it, "inertia: the property of a body in virtue of which it tends to persist in a state of rest or uniform motion in a straight line" (Pitt 1977: 193). Clearly, inertia is a concept that figures in causal explanation.

Further, to deny that it is commits one to causal indeterminism in surprising circumstances. Causal determinism says that every event has an immediately prior cause. But a spaceship moving in the absence of forces has no cause, if we deny persistence as causation; so causal determinism fails.

Immanent causation is also required for the causal theory of identity. Opponents of immanent causation, and the above account of Newton's First Law, do not generally appear to be aware of this fact. As Armstrong (as mentioned earlier) and Russell (see Chapter 4) have demonstrated, the notion of immanent causality can be used to explain our sense that things have identity over time. I do not accept the causal theory of identity, but it is a popular view, and opponents of immanent causation should be aware that they thereby rule out this popular view.

Armstrong has suggested (in conversation) that we think of persistence as a sort of transference – transfer of energy/momentum from an earlier part to a later part of an object. I have no serious objection to this, although it does seem odd that the quantity is said to be transferred when in fact it is retained (in effect, this suggestion denies that there is such a thing as retaining a quantity). But it will not save the transference theory, which requires that the quantity be transferred *from one object to another*.[11] In fact, the right kind of modification to the transference theory amounts to an admission that there is another kind of causation besides transference.[12]

the following questions. In the case of the moving spaceship, is something happening? Is that happening uncaused? If not, what is the cause?
11. If we think of objects as worldlines – timelike worms – and think of the transference from one timeslice to another, we still have the problem that the energy hasn't been given a different object in the relevant sense. For transference to occur, the object that gains mustn't have had it before, and the object that gives it up must no longer have it. This is essential to the meaning of 'giving,' and 'transference.'
12. The same objection applies to the 'force' theories of Bigelow, Ellis and Pargetter (1988), Bigelow and Pargetter (1990a; 1990b), and Heathcote (1989).

Of course, none of this entails that there is no such thing as causation in the sense of transference; just that there is another type of causation – causation as persistence. As well as the passing on of the appropriate quantity, the mere *possession* of the quantity involves an object in causality. So we could give an account of causation as giving *or* keeping, that is, transference *or* persistence. Then, in our world the thing that is possessed, and transferred, turns out to be energy or momentum. The transference analysis becomes the Spinozean Disjunction.[13]

III.4.2 Identity of Physical Quantities

A second problem concerns the identification and reidentification of quantities like energy and momentum. Quine, in his discussion, takes it as a necessary assumption of the transference approach that energy be pictured as being like matter in the sense that it is traceable from point to point through time (1973: 5). Aronson makes a detailed case for recognising the identity through time that such quantities possess (1971a: 143–149). He argues that "words such as 'transfer,' 'transform,' 'convert,' 'exchange' etc, [are] words whose application usually involves some kind of identity through time" (1971a: 143, note 13). Further, the very concept of conservation laws presupposes that those quantities retain their identity through time. If, for example, energy were annihilated and recreated in an interaction, then the conservation laws would be violated. Rather, energy is converted or transferred. "Denial of numerical identity of quantities is incompatible with conservation principles" (1971a: 149). Fair uses the conservation laws as a technique for identifying energy possessed by an object at a given time as being (part of) the same energy that was possessed by another object at an earlier time. Thus, "identity conditions for energy and momentum are singularly perspicuous among the class of properties in general" (1979: 234).

But what sort of thing is the quantity, energy, which is possessed by an object, yet has an identity through time? 'Possesses' is ambiguous. When we say "x possesses y" we could mean 'instantiates' – where an individual possesses a property, such as in "she possesses a warm smile" (like "the chair is red"). Or we could mean that x and y are two indi-

---

13. In Chapter 5, section 5.2, we consider in more detail the connection between identity and causation.

viduals that stand in a certain range of asymmetric relations such as ownership, or some sort of physical control, as in "I possess that book." In other words, 'possesses x' can be a one-place or a two-place predicate. The question is, which is 'possesses energy'?

One line of thought leads us to suppose it is the former, that is, that energy is a property, a quantitative property that individuals instantiate. Energy or momentum, while capable of being possessed by different objects consecutively, do not exist uninstantiated, that is, possessed by nothing (Aronson 1971a: 145). So energy seems to be a quantitative property. But as a quantitative property, it has a peculiar characteristic: according to Aronson and Fair, it has an identity through time.

What could the Aronson-Fair notion of identity amount to? It cannot be the sort of identity that properties usually exhibit (two objects have the same property in the sense that they instantiate the one universal; this chair and that book are both the same colour). When Aronson and Fair speak of objects having the same energy, they do not mean it in this sense (although energy is a universal that objects share). Nor can identity mean 'the same amount,' although this is a straightforward characteristic of quantitative properties, when there is a numerical equality between two instances. In the case of energy and momentum, the conservation laws demand this numerical equality through time for any closed system. But again, Aronson and Fair mean more than this – not only is there the same amount of energy, but there is the same stuff existing over time.

This brings us to a third sense of identity, sometimes called genidentity (see Grünbaum 1973: 189) – identity of substance or matter over time. Objects, matter, substance, individuals all have an identity over time that enables us to say that this is the same thing as was here at an earlier time. Presumably persons have this sort of identity through time. This is not a property in the usual sense, nor is it obviously a relation between two objects. Clearly, it is this third sense of identity that Aronson and Fair have in mind. But how can a *property* have identity over time? We don't normally think of properties as having that sort of identity. We don't say that the red of this book is the same red as the book had an hour ago, unless we mean it in the first sense of identity. Similarly for quantitative properties – we don't say that the tree still has the same height unless we mean identity in the second sense. In fact, the notion of identity over time leads us to think that energy is not a property at all, but a thing, a substance, separate from the object, but standing in some relation to it (that is, the second sense of 'possesses'). Thus the notion of the identity of energy is not at all clear.[14]

To summarise, Aronson and Fair claim that energy is a quantity that cannot be unpossessed and that has identity through time. But this presents them with a dilemma. If it is a substance, possessed in the sense of a relation like ownership, then it can exist unpossessed. But if it is a quantitative property, possessed in the sense of 'instantiated,' then it cannot have identity through time.

Consider as an analogy, possessing money. If we are thinking of tokens – notes and coins – then when I deposit and withdraw money from my account I can have the same money, in two senses. There can be a numerical equality, if I've been given back the same amount; and there can be an identity through time, if I've been given back exactly the same notes and coins as I had deposited. But in this latter sense – identity over time – we are talking about objects, not a property. Such money (i.e., the tokens) can be possessed by no one. On the other hand, if we are thinking about money in an electronic sense, then, when I deposit and withdraw from my account I can be left with the same money only in the sense of numerical equality. In that case, the money I possess is a property of me, and money in that sense cannot be possessed by no one. But then there is no identity through time – it makes no sense to speak of getting the same money back in that sense.

Aronson uses the analogy of American football, where the ball is possessed by either team but never by neither team (1971a: 145, note 16). But this is ambiguous in roughly the same way that 'possessing money' is ambiguous. We could mean 'possesses the ball' in some physical sense: one of the players is holding it. In that sense the ball has an identity over time, but then 'possesses the ball' is really a relation to an object, not a nonrelational property. But in this sense neither team possesses the ball when it is lying on the ground between plays. On the other hand, we could mean by 'possesses the ball' the formal sense defined by the rules of the game, in which case it cannot be possessed by neither team. But in this sense there is no identity over time. Having the same ball from play to play is not a formal part of the game – you could swap balls (tokens) during a time-out without making a differ-

14. In his recent book, Ehring (1998) avoids this dilemma by identifying causation with trope persistence, where tropes are understood to be persisting abstract particulars that are passed from one object to another. See also Kistler (1998).

ence to the game. The ball in the formal sense doesn't have identity through time, although its physical tokens do.

Indeed, it seems that there are no complete analogies to be had. One analogy that we actually use for energy is water – we speak of energy flowing, and we envisage something like water, or some liquid substance. But water is not a quantitative property with identity over time. It is an independent object possessed only in the sense of a relation. It can stand alone, 'uninstantiated.' There are problems with reidentifying portions of water, much like the problems we will shortly consider for energy, but for water this is just a pragmatic, computational problem. In reality its molecules have identity through time, as genuinely as any object.

One way of clearing the conceptual mess is to deny that energy and momentum have this sort of identity. This is suggested by Dieks (1986: 88), who argues that the scientific understanding of energy is that it is conserved in a global sense, but that the notion that the same energy is given from one object to another is empirically unverifiable in classical physics, and inconsistent with quantum mechanics. It would be consistent with Dieks's suggestion to suppose that energy and momentum are quantitative properties that have no identity that can be transmitted through time. Then talk of 'transfer' and 'exchange' of energy is to some extent metaphorical.

This, if true, would explain why we run into indeterminancies in many-body problems. Fair gives the example (1979: 238) of three oscillating devices, A, B and C, on a pond. There is a point, p, where the waves created by A and B interfere constructively, but where the waves created by C interfere destructively. The net effect at that point is as if just one of A and B is operating, and C is not operating. Suppose there is a float on the water at p. Then whose energy, A's or B's, causes it to bob up and down? Fair suggests that it is indeterminate, and that causation is also.

A similar case is presented by Ehring (1986: 256):[15] suppose part of A's energy is given to B, and then part of B's energy is given to C. Was A's energy given to C? The answer is that mere conservation laws simply do not tell us. Or consider four hanging balls. Two swing towards the other two, which are stationary and touching. They all collide at the same instant and the collision leaves the first two stationary and touch-

---

15. Although for some reason Ehring seems to think that his case differs from Fair's. It's also curious that Ehring mistakenly credits Quine with making this point.

ing, while the other two move off. Which motion was given to which ball? It seems that whenever we move to many-body problems the conservation laws leave the question of the identity over time of the energy indeterminate. Now, it could be just a hidden fact – both which energy goes where, and what actually causes what. But, curiously, Fair says that it is indeterminate, with respect to both energy transfer and causation. Given that this sort of indeterminacy is so widespread, this seems to entail that energy generally doesn't have identity through time.

The argument put by Aronson and Fair is that the laws of conservation *require* this idea of identity. But we can see now that this argument fails. The conservation laws require a numerical equality. Perhaps it makes some sense, in the case of a single body passing on energy to another body, to think of the energy gained as genidentical to the energy lost. But when we consider many-body problems we recognise that genidentity is not determined by the conservation laws, and it no longer makes much sense to think that it is required. Therefore, regardless of who is right about the identity over time of physical quantities, that kind of identity is not entailed by the truth of conservation laws.

So the transference theory has a problem. It seems to be very difficult to make sense of the genidentity of a quantitative property of the sort utilised in the transference theory. Indeed, it is apparent that the notion of causation as 'giving' is to some extent misleading.

III.4.3 Causal Asymmetry

We will address the question of causal asymmetry in detail in Chapter 8. However, we need to mention a relevant difficulty for Aronson's version of the transference account.[16] According to Aronson, the direction of transference gives the causal asymmetry, but the direction of transference is independent of the direction of time. If energy were transferred from a later event to an earlier event, then we would have a case of backwards-in-time causation. And, according to Aronson, causal asymmetry is an area where the transference account scores considerably better than the rival regularity or manipulability accounts.

However, it can be shown that Aronson's account theory fails to

---

16. A number of authors have pointed to difficulties that transference theories have with the notion of causal asymmetry. The first were Earman (1976: 24–25) and Beauchamp and Rosenberg (1981: 210–211).

deliver the goods. First, there is an epistemological problem. How do we know that the quantity was transferred from the earlier event A to the later event B? How do we know that it wasn't transferred from B to A? We cannot say that we take the earlier event to be the cause, because the account is supposed to provide the direction independently of time. Nor can we say that the conservation laws tell us – they are symmetrical with respect to time.

There is an equivalent ontological problem as well. What does it mean to say that energy is transferred *in a direction*? Aronson does not say. He cannot say that it means at the earlier time A possessed it and B didn't; and then at a later time B possessed it and A didn't, because that is to appeal to temporal priority. The direction of transference cannot be specified by appeal to physics, either, for physics is time-symmetric. But what else could Aronson mean? That he has completely failed to specify certainly does undermine his claim to have satisfactorily dealt with the issue!

As an illustration of the problem, let us ask, what is it for something to be given, as opposed to received? In particular, what is it for something to be given, forwards in time as opposed to being received, backwards-in-time? For example, suppose I give you a book. Without appealing to the direction of time, the only facts we have are that at one point in time I have the book and you don't, and that at another point in time you have the book and I don't. But these facts are completely time-symmetric. So Aronson claims to have given an account of the direction of causation in terms of transference, without appealing to temporal priority, but in fact has failed to deliver the goods.

Notice, further, that if there is a problem for the contingent identity thesis then it is equally a problem for the meaning analysis that I considered earlier in this chapter. It's all very well to say "causation is transference," but if 'transference' itself is obscure, we may as well have stayed with 'causation.'

III.5 SUMMARY

I have shown (see section 3.2) that the transference theory is well equipped to handle indeterministic causation, and so is not open to the kind of objections that were levelled in Chapter 2 against the regularity and probabilistic theories. Further, I have considered a way to extend the transference theories so as to provide a meaning analysis as well as a contingent hypothesis about the way our world is. Transference theorists claim that in our world causation amounts at bottom to the transfer of energy or momentum. They do not claim that we mean anything about energy when we speak of causation. The extension is to suppose that the idea of transfer, or giving, in general is a core feature of what we *mean* in our use of causal language. Then we may suppose that it has been discovered that in our world the thing that is transferred is energy or momentum. Thus we have come upon a new and comprehensive approach to causation.

However, there are problems with the transference account. Unfortunately, the core notion of 'transfer,' in particular in relation to the direction of transfer and the identity over time of the transferred quantities, turns out to be most obscure, if not incoherent.

In addition, there appears to be a type of causation, in our world, that is ruled out by the transference theory. This is the case of persistence as causation. The example of a spaceship's inertia being the cause of its motion is a good example. We need to be able to incorporate such instances of causation.

Nevertheless, there is important insight in the transference theory. In what follows I offer a way to salvage the valuable parts of the account, without being committed to the dubious parts.

4

Process Theories of Causation

We now turn to another promising line of investigation – the notion of a causal process. We begin with a discussion of Bertrand Russell's notion of a 'causal line,' which, I claim, captures what is lacking in transference theories: immanent causality. However, it fails to deal with the important distinction found in special relativity between causal and pseudo processes. Thus we turn, in section 4.2, to Wesley Salmon's theory of causal processes and interactions, according to which the causal connections in the world are best analysed in terms of causal processes and interactions, where causal processes are those capable of transmitting 'marks.' The former part of this is defended, the latter part rejected. It is argued (section 4.3) that there are four areas of inadequacy with regard to Salmon's theory: that the theory is circular, that the mark theory is extensionally inadequate, that statistical forks do not serve their intended purpose, and that Salmon does not achieve his stated desire to avoid 'hidden powers.' Nevertheless, Salmon's account is an improvement in important respects on the regularity, probabilistic and transference accounts.

IV.1 RUSSELL'S CAUSAL LINES

Bertrand Russell is famous for the view, presented in 1913, that the philosophers' concept of 'cause' is otiose in modern science.[1] But later in life, in *Human Knowledge* (1948: 453–460), Russell offers a quite different account. He still regards 'cause,' as used by philosophers, to be otiose but now thinks that advanced science uses causal notions that have been developed from the primitive concept.

In *Human Knowledge*, Russell introduces a notion of 'quasi-permanence,' by which he means the sort of similarity that things

---

1. For a different reading of Russell, see Dowe (1997b).

display over time – not an absolute similarity, but a rough similarity. Quasi-permanence is displayed by the 'things' of common sense, which normally don't change discontinuously. and by the physical matter of classical physics; and those items that possess quasi-permanence Russell calls 'causal lines' – "more or less self determined causal processes" (1948: 459). Russell describes causal lines thus:

A causal line may always be regarded as the persistence of something – a person, a table, a photon, or what not. Throughout a given causal line, there may be constancy of quality, constancy of structure, or gradual change in either, but not sudden change of any considerable magnitude. I should consider the process from speaker to listener in broadcasting one causal line; here the beginning and end are similar in quality as well as structure, but the intermediate links – sound waves, electromagnetic waves, and physiological processes – have only a resemblance of structure to each other and to the initial and final terms of the series. (1948: 459)

An example of a causal line, Russell says, is an object moving according to Newton's First Law of Motion, which states that an object will continue in motion unless acted on by a force. This implies that the object will continue to exist, and continue to have the relevant properties in a continuous fashion. Russell takes this as an example of quasi-permanence (1948: 457–458).

Russell's purpose in developing these ideas is to show how scientific inferences are possible. The problem with thinking about causal laws as the underpinning of scientific inference is that the world is a complex place, and while causal laws may hold true, they often do not obtain because of preventing circumstances, and it is impractical to bring in innumerable 'unless' clauses. But, even though there is infinite complexity in the world, there are also causal lines of quasi-permanence, and these warrant our inferences.

Russell elaborates these ideas into five postulates that he says are necessary "to validate scientific method" (Russell 1948: 487). The first is The Postulate of Quasi-permanence, which states that there is a certain kind of persistence in the world, for generally things do not change discontinuously. The second postulate, Of Separable Causal Lines, allows that there is often long-term persistence in things and processes. The third postulate, Of Spatio-temporal Continuity, denies action at a distance. Russell claims that "when there is a causal connection between two events that are not contiguous, there must be intermediate links in the causal chain such that each is contiguous to the next, or (alternatively) such that there is a process which is continuous" (1948: 491). The Structural Postulate, the fourth, allows us to infer from structurally similar complex events ranged about a centre to an event of similar structure linked by causal lines to each event. The fifth postulate, Of Analogy, allows us to infer the existence of a causal effect when it is unobservable.

Here, in Russell's idea of a causal line, we have the sort of notion that we sought in the previous chapter (section 3.4), where we considered the case of a spaceship moving beyond the influence of any field: its inertia should be regarded as the cause of its continuing motion; and more generally, any object persisting over time can be thought of as a causal process. We concluded that transference theory is unable to accommodate such cases of causation as persistence, but Russell's causal lines seem to be just the right sort of thing.

Nevertheless, there are problems with Russell's formulation. Wesley Salmon has pointed to two difficulties (1984: 140–145). First, and most significantly, Russell does not distinguish between causal and pseudo processes, a distinction first noticed by Hans Reichenbach (1958: 147–149; first published in German in 1928) as he reflected on the implications of Einstein's Special Theory of Relativity. Reichenbach noticed that the central principle that nothing travels faster than the speed of light is 'violated' by certain processes. Examples include a spot of light moving along a wall, which is capable of moving faster than the speed of light (one just needs a powerful enough light and a wall sufficiently large and sufficiently distant), a moving shadow, and the point of intersection of two rulers.[2] Such pseudo processes (as I shall call them)[4] do not refute special relativity, Reichenbach argued, simply because they are not causal processes, and the principle that nothing travels faster than the speed of light applies only to causal processes (1958: 204). Thus special relativity demands a distinction between causal and pseudo processes, but Russell's characterisation fails to rule out pseudo processes, for these also display similarity of quality and similarity of structure.

We met a related problem in the previous chapter, in comparing Aronson's and Fair's versions of the transference theory. Aronson allows that velocity may be the quantity transferred; but his version is open to the objection of the two coloured spots that appear to interact, and which by Aronson's criteria do transfer velocity. Fair's ver-

---

2. See Salmon's clear exposition (1984: 141–144).

sion is not open to that difficulty, because he allows only energy and momentum as the appropriate quantities. We can now see that what is wrong with the 'interacting' spots is that they are pseudo processes, not causal processes.[3]

Salmon's second objection is that Russell's account is formulated in terms of how we make inferences. For example, Russell says,

A "causal line," as I wish to define the term, is a temporal series of events so related that, given some of them, something can be inferred about the others whatever may be happening elsewhere. (1948: 459)

Salmon's criticism of this is precisely that it is formulated in epistemic terms, "for the existence of the vast majority of causal processes in the history of the universe is quite independent of human knowers" (1984: 145). Salmon, as we shall see in the next section, develops *his* account of causal processes as an explicitly 'ontic,' as opposed to 'epistemic,' account (see 1984: chap. 1).

There is a further reason why Russell's epistemic approach is unacceptable. While it is true that causal processes *do* warrant inferences of the sort Russell has in mind, it is not the case that all rational inferences are warranted by the existence ('postulation,' in Russell's thinking) of causal lines. There are other types of causal structures besides a causal line. Russell himself gives an example: two clouds of incandescent gas of a given element emit the same spectral lines, but are not causally connected (1948: 455). Yet we may rightly make inferences from one to the other. A pervasive type of case is where two events are not directly causally connected but have a common cause.[5]

In addition to these two criticisms levelled by Salmon, there are two further weaknesses in Russell's account of causal lines. One is that he wants to explicate causal lines in terms of causal relations, at base: a causal line is a temporal series of events, all related in turn as cause and effect. Thus an object is "not to be regarded as a single persistent substantial entity, but as a string of events having a certain kind of causal connection with each other" (1948: 458). But if we reduce causal processes to causal relations, then we are back with the problem of saying what the causal relation is, and with the problem that we may

3. Reichenbach called them 'unreal sequences' (1958: 147–149).
4. In the next chapter I shall argue that quantities such as velocity do not serve to characterise causal processes, whereas quantities like energy and momentum do.
5. Explicated by Reichenbach and Salmon (see the following section).

have again denied causation as persistence. In the next section we will see that in developing his account Salmon resolves to keep causal processes as basic. I think that is the right approach.

The second weakness is that, while an account of causal lines does apparently (the previous point notwithstanding) account for causality as persistence, it does so at the cost of denying causality as transference. For if causation is to be explicated in terms of causal lines, and if these concern causation as persistence, then there is no place left for the interaction between one process and another. Now, in fact, Russell does not say that causal lines are the full story on causation, but if they were then there would be no room for transference. This would be the opposite mistake to that of the transference theory. What we need, as we saw in the last chapter, is an account that allows for *both* transference and persistence – the Spinozean Disjunction. It certainly won't do to abandon transference in order to have persistence.

IV.2 SALMON'S PROCESS THEORY OF CAUSALITY

Wesley Salmon's theory of causality, developed in a series of articles spanning twenty years and systematically presented in his book *Scientific Explanation and the Causal Structure of the World* (1984), has been highly influential amongst philosophers and scientists.[6] (Salmon's more recent views will be considered in Chapter 5.) The theory treats causality primarily as a property of individual processes and can for this reason be termed a 'process' theory. However, although he draws on the work of Reichenbach and Russell, Salmon's theory is highly original and contains many innovative contributions. His broad objective is to offer a theory that is consistent with the following assumptions:

(i) causality is an *objective* feature of the world;
(ii) causality is a *contingent* feature of the world;
(iii) a theory of causality must be consistent with the possibility of *indeterminism*;
(iv) the theory should be (in principle) *time-independent* so that it is consistent with a causal theory of time;
(v) the theory should not violate Hume's strictures concerning 'hidden powers.'

6. For example, important elements have been reprinted in various collections, such as Pitt (1988) and Sosa and Tooley (1993).

Salmon proposes to overcome traditional difficulties with determining the nature of the causal relation by treating causality as primarily a characteristic of continuous processes rather than as a relation between events. The theory involves two elements, the production and the propagation of causal influence (1984: 139).

The latter is achieved by *causal processes*. Salmon defines a process as anything that displays consistency of structure over time (1984: 144). To distinguish between causal and pseudo processes, Salmon makes use of Reichenbach's 'mark criterion': a process is causal if it is capable of transmitting a local modification in structure (a 'mark') (1984: 147). Drawing again on the work of Bertrand Russell, Salmon seeks to explicate the notion of 'transmission' by the "at-at theory" of mark transmission. The principle of mark transmission (MT) states:

MT: Let P be a process that, in the absence of interactions with other processes would remain uniform with respect to a characteristic Q, which it would manifest consistently over an interval that includes both of the space-time points A and B ($A \neq B$). Then, a mark (consisting of a modification of Q into Q'), which has been introduced into process P by means of a single local interaction at a point A, is transmitted to point B if [and only if] P manifests the modification Q' at B and at all stages of the process between A and B without additional interactions. (1984: 148)

(See the later discussion for the precise characterisation of 'interactions.') Salmon himself omits the 'only if' condition, but, as suggested by Sober (1987: 253), this condition is essential, because the principle is to be used to identify pseudo processes on the grounds that they do not transmit a mark.[7] Thus for Salmon a causal process is one that can transmit a mark, and it is these spatiotemporally continuous processes that propagate causal influence.

For example, a baseball moving through the air is a causal process, since it can be marked – for example, by making a small cut with a knife – and since the mark would be transmitted because the modification would continue to exist at every spacetime point on the ball's trajectory, provided there is no further interaction. On the other hand, a spot of light moving across a wall cannot be marked by a single local

---

7. But we might note that Sober's claim that MT does not provide a necessary condition for mark transmission because marks can be transmitted when many processes are involved (1987: 253) misreads the principle MT, which merely claims: a mark introduced *by a single interaction* is transmitted if and only if that mark appears at every point.

modification. If you do something to the spot – for example, change its shape by distorting the surface – that 'modification' is not subsequently transmitted and so doesn't count as a mark.

To accompany this theory of the propagation of causal influence, Salmon also analyses the production of causal processes. According to Salmon, causal production can be explained in terms of causal forks, whose main role is the part they play in the production of order and structure of causal processes, and which are characterised by statistical forks. To Reichenbach's 'conjunctive fork' Salmon adds the 'interactive' and 'perfect' forks, each corresponding to a different type of common cause.

The 'conjunctive fork' (Reichenbach 1991), where two processes arise from a special set of background conditions, often in a nonlawful fashion (Salmon 1984: 179), involves a statistical correlation between the two processes that can be explained by appealing to the common cause, which 'screens off' the statistical connection. Stated formally (for the full account see 1984: chap. 6), three events A, B and C form a conjunctive fork if:

$$P(A.B) > P(A).P(B) \tag{1}$$

and

$$P(A.B|C) = P(A|C).P(B|C) \tag{2}$$

The principle of the common cause says that where we have two events, A and B, where (1) holds, and there is no direct causal connection between A and B, then we seek to explain this correlation by finding a third earlier event C such that (2) holds. In Salmon's theory of causality, conjunctive forks produce structure and order from 'de facto' background conditions (1984: 179).

Second, there is the 'interactive fork,' where an intersection between two processes produces a modification in both (1984: 170) and an ensuing correlation between the two processes cannot be screened off by the common cause. Instead, the interaction is governed by conservation laws. Salmon calls this a *causal interaction*. For example, consider a pool table where the cue ball is placed in such a position relative to the eight ball that, if the eight ball is sunk in one pocket A, the cue ball will almost certainly drop into another pocket B. There is a correlation between A and B such that equation (1) holds. But the common cause C, the striking of the cue ball, does not screen off this correlation. Salmon suggests that the interactive fork can be characterised by the relation

$$P(A.B|C) > P(A|C).P(B|C) \qquad (3)$$

together with (1) (1978: 704, note 31). By 'interactive fork' we will mean just 'a set of three events related according to equations (1) and (3).' Salmon's idea is that interactive forks are involved in the production of modifications in order and structure of causal processes (1982b: 265; 1984: 179).

The idea of a causal interaction is further analysed by Salmon in terms of the notion of mutual modification. The principle of causal interaction (*CI*) states:

CI: Let P_1 and P_2 be two processes that intersect with one another at the spacetime S, which belongs to the histories of both. Let Q be a characteristic that process P_1 would exhibit throughout an interval (which includes subintervals on both sides of S in the history of P_1) if the intersection with P_2 did not occur; let R be a characteristic that process P_2 would exhibit throughout an interval (which includes subintervals on both sides of S in the history of P_2) if the intersection with P_1 did not occur. Then, the intersection of P_1 and P_2 at S constitutes a causal interaction if (1) P_1 exhibits the characteristic Q before S, but it exhibits a modified characteristic Q' throughout an interval immediately following S; and (2) P_2 exhibits R before S but it exhibits a modified characteristic R' throughout an interval immediately following S. (Salmon 1984: 171)[8]

Third, the deterministic limit, where $P(A.B|C) = 1$, needs to be treated separately, because it may represent a limiting case of either a conjunctive or interactive fork. Thus it is impossible to tell by statistical consideration whether the case in hand is a conjunctive or interactive fork. Salmon (1984: 177–178) proposes to call such cases 'perfect forks,' to draw attention to this difficulty.

Salmon usually uses the words 'order' and 'structure' with reference to the causal 'thing' that is produced, but some clarification is needed of this usage. First, on 'structure,' Salmon writes, "A given process,

---

8. Sober (1987: 255) is mistaken in drawing a similar conclusion for *CI* as he did for *MT* (see the previous discussion) – namely, that Salmon needs it to read as a necessary condition – for the simple reason that not every characteristic of the two processes need be modified by an interaction.

whether it be causal or pseudo, has a certain degree of uniformity – we may say, somewhat loosely, that it exhibits a certain structure" (1984: 144). As Salmon quotes Russell's "constancy of quality, constancy of structure" description of causal lines, where Russell is describing a uniformity or sameness over time, it seems likely that Salmon means something similar (Russell 1948: 459; 1984: 144, 153). Thus Salmon speaks of processes exhibiting 'regularity' (1984: 144–145). This 'structure' or uniformity is produced by conjunctive forks, modified by interactive forks, and transmitted by causal processes, and is closely related to the transmission of 'energy' and 'information' (1984: 154–155, 261).

On the other hand, 'order' seems to be a correlation between processes, or among many processes. For example, Salmon compares the conjunctive fork to Reichenbach's hypothesis of the branch structure, which describes how order arises from preexisting order, for example, how low entropy states arise from available energy in the environment (1984: 180). 'Order' then has to do with the correlations of processes, or the regularity of many processes, but 'structure' refers to enduring qualities or properties of an individual process. Both are produced by causal forks and propagated by causal processes.

It should be noted that the statistical relations so central to most probabilistic theories of causality here only characterise production mechanisms. In fact, the term 'probabilistic' refers to the fact that an individual causal process may have a less-than-deterministic tendency to produce a certain interaction (1984: 202) rather than to the existence of a relation of positive statistical relevance. The strength of this tendency Salmon calls 'propensity.' Propensities are probabilistic causes, produced by interactions and transmitted by causal processes (1984: 204). These propensities generate frequency data (1988a: 25), such as the statistical relations mentioned earlier, but themselves apply to the single-case. Thus, in Salmon's 'process' theory of causality, (indeterministic) causality is more fundamental than statistical relations. Indeed, Kitcher has suggested that Salmon's theory be termed 'indeterministic causality' rather than 'probabilistic causality' (Salmon 1988b: 321).

Finally, it was noted earlier that one of Salmon's broad objectives is that causality be understood as 'contingent.' It is important to clarify what is meant by this. Salmon does *not* regard the task as one of providing a conceptual analysis applicable to all logically possible worlds

(1984: 239–242). Rather, Salmon's theory is aimed at providing a description of contingent features in the actual world, such as the fact that conjunctive forks are open only to the future (1984: 164, 240), the fact that there are records of the past but not of the future (1984: 72), and the fact that macroscopic processes are spatiotemporally continuous (1984: 241–242). These features are contingent, de facto, a posteriori facts about the actual world. Many philosophers would disagree with this approach; those who have done so explicitly include Fetzer (1987: 609) and Hanna (1986: 584). Others base counterarguments on the assumption that Salmon is attempting a conceptual analysis. Tooley, for example, argues against Salmon's spatiotemporal continuity requirement and his account of causal priority by presenting logically possible nonactual situations where these features are not present (1987: 235–238). But this is to misunderstand Salmon's broad desiderata. (See Chapter 1, this volume; and Dowe 1989). Salmon's theory must be taken as an attempt to understand the workings and nature of the actual world.

To summarise, the key notions in Salmon's account are causal processes and causal interactions, both of which are understood in terms of mark transmission. These explications can be summarised in five propositions.[9] (Here some of the counterfactual formulation is omitted because it is irrelevant to the current discussion.)[10]

I. A process is something that displays consistency of characteristics.
II. A causal process is a process that can transmit a mark.
III. A mark is transmitted over an interval when it appears at each spacetime point of that interval, in the absence of interactions.
IV. A mark is an alteration to a characteristic, introduced by a single local interaction.
V. An interaction is an intersection of two processes where both processes are marked and the mark in each process is transmitted beyond the locus of the intersection.

9. Originally given in Dowe (1992c: 200), where I thought an extra proposition would be needed to account for the difference between a causal interaction and any old interaction. However, I now realise that Salmon counts all interactions as causal (see 1994: 299). Further, a clause was omitted in that first analysis that now appears in V, to the effect that the mark is to some extent permanent.
10. The counterfactual version of these propositions will be presented and discussed in section 4.3.

IV.3.1 Circularity

Some commentators have claimed to find circularity in Salmon's analysis of causality. Kitcher, Mellor, and Sober have each, in different ways, laid this charge (Kitcher 1985; Mellor 1988; Sober 1987). These critiques are brief, but I think they point us to a real difficulty, namely, the possibility that the definitions of 'mark' and 'interaction' might be mutually dependent, rendering the account circular. That is, are IV and V circular? It seems that they are.

The concept of a mark involves the concept of an interaction: a mark is a modification to a process introduced by a single interaction (IV). But the concept of an interaction involves the concept of a mark: an interaction is an intersection where both processes are marked (V). The circularity is blatant. In short, the concepts of 'mark' and 'interaction' are mutually dependent, so I conclude that the account is unacceptably circular.

In a recent reply to this analysis, Salmon (1994: 298–299) replaces 'interaction' in IV with 'intersection.' He replaces IV with:

S-II. A mark is an alteration to a characteristic that occurs in a single local intersection.

The first comment to make about this is that this does not square with Salmon's original *MT* principle (quoted earlier). That principle speaks of "a mark (consisting of a modification of Q into Q'), which has been introduced into process P by means of a single local *interaction* at a point A" (my italics) (1984: 148). So what Salmon proposes in S-II must be taken as a correction to the *MT* principle, not to my analysis.

Is this an acceptable correction? One might think not, for the following reason. If a mark can be a change in a characteristic of a process arising in a single local *intersection*, then there's nothing stopping accidental correlation between a random change in a process and an intersection with another process. For example, two spots intersect on a screen at the same instant as a filter is placed on one of the spots (at its source). Then the change of colour will count as a mark, and so the moving spot will count as a causal process.

However, Salmon has another clause in his *MT* principle, which is not included in the formulation I–V; namely, the counterfactual condition that the process would remain uniform with respect to that characteristic, if the mark had not occurred. This is not true of the case mentioned here. For the spot that is changed to red would still have changed to red in the absence of the intersection with the other process. We will examine this counterfactual clause later in this chapter; but at this stage we should simply note that if Salmon is allowed that clause, then the change from 'interaction' to 'intersection' does remove the circularity in a satisfactory manner.

However, there is a second point of circularity that looms; this time concerning the interrelation between the concepts of 'mark transmission' and 'interaction,' in III and V respectively. For the concept of transmission involves the concept of an interaction: transmission must be in the absence of interactions (III). But the concept of an interaction involves the concept of transmission: in an interaction both processes transmit a mark beyond the locus of the intersection (V). Again, the circularity is blatant.

First, it must be noted that the concept of transmission cannot be removed from V, even though that would remove the circularity. For if it were, one would be able to mark a pseudo process. For example, holding up a red filter near the wall changes a characteristic (the colour) of the moving spot, if only momentarily. Such a change will not persist, and that's really the hallmark of a pseudo process: any attempted mark will not be permanent. So the clause that insists that the mark be transmitted beyond the locus of the intersection must be retained.

The alternative strategy would be to replace 'interaction' in III with 'intersection.' The requirement for a mark to be transmitted would then be that the mark appears at each spacetime point of the interval in the absence of intersections with other processes. Such a requirement would satisfactorily rule out every undesirable case that its predecessor did, simply because if a process moves in the absence of further interactions then it must by definition move in the absence of intersections. However, the worry would be that the condition rules out much more besides. In fact, this change entails that many causal interactions will not qualify as such by V because the changes produced in the processes are not transmitted in the absence of further intersections with other processes. For example, a perfectly good interaction between two billiard balls will not qualify as a interaction because the path of the ball soon after the collision happens to intersect with the path of the shadow of a moth, say.

Points of inadequacy of this sort will be discussed in the next section, and it should be said that this change will only worsen the situation, as we shall see. Nevertheless, the circularity is removed. Which poison Salmon chooses is his prerogative. For the purposes of the following discussion we will continue to treat the account given at the start of this section, in propositions I to V, and in particular, we will continue to speak of interactions.

IV.3.2 The Adequacy of the Mark Theory

The mark transmission (*MT*) principle carries a considerable burden in Salmon's account, for it provides the criterion for distinguishing causal from pseudo processes. However, it has serious shortcomings in this regard. In fact, it fails on two counts: it excludes many causal processes; and it fails to exclude many pseudo processes. We shall consider each of these problems in turn.

1. *MT excludes causal processes.* First, the principle requires that processes display *a degree of uniformity over a time period*. This distinguishes processes (causal and pseudo) from 'spatiotemporal junk,' to use Kitcher's term. One problem with this is that it seems to exclude many causal effects that are short-lived. For example, short-lived sub-atomic particles play important causal roles, but they don't seem to qualify as causal processes. On any criterion there are causal processes which are 'relatively short-lived.' Also, the question concerning how long a regularity must persist raises philosophical difficulties about degrees that need answering before we have an adequate distinction between processes and spatiotemporal junk. However, if these were the only difficulties I think that the theory could be saved. Unfortunately, they are not.

More seriously, the *MT* principle requires that causal processes remain uniform *in the absence of interactions* and that marks can be transmitted *in the absence of additional interventions*. In real situations, however, processes are continuously involved in interactions of one sort or another (1989: 464). Even in the most idealised of situations interactions of sorts occur. For example, consider a universe that contains only one single moving particle. Not even this process moves in the absence of interactions, for the particle is forever intersecting with spatial regions. If we require that the interactions be causal (at the risk of circularity), then it is still true that in real

74

cases there are many causal interactions continuously affecting processes. Even in carefully controlled scientific experiments there are many (admittedly irrelevant) causal interactions going on. Further, Salmon's central insight that causal processes are self-propagating is not entirely well-founded. For while some causal processes (light radiation, inertial motion) are self-propagating, others are not. Falling bodies and electric currents are moved by their respective fields. (In particular, there is no electrical counterpart to inertia.) Sound waves are propagated within a medium, and simply do not exist 'in the absence of interactions.' Such processes require a 'causal background,' and some can even be described as being a series of causal interactions. These causal processes *cannot* move in the absence of interactions. Thus there is a whole range of causal processes that are excluded by the requirement that they remain uniform in the absence of any interactions.

It seems desirable, therefore, to abandon the requirement that a causal process be one that is capable of transmitting a mark in the absence of further interactions. However, the requirement is there for a reason, and that is that without it the theory is open to the objection that certain pseudo processes will count as being capable of transmitting marks. Salmon considers a case where a moving spot is marked by a red filter held up close to the wall. If someone runs alongside the wall holding up the filter, then it seems that the modification to the process is transmitted beyond the spacetime locality of the original marking interaction. Thus there are problems if the requirement is kept, and there are problems if it is omitted. So it is not clear how the theory can be saved from the problem that some causal processes cannot move in the absence of further interactions.

2. *MT fails to exclude pseudo processes.* Salmon's explicit intention in employing the *MT* principle is to show how pseudo processes are different from causal processes. If *MT* fails here, then it fails its major test. However, a strong case can be made for saying that it does indeed fail this test.

First, there are cases where pseudo processes qualify as being capable of transmitting a mark, because of the *vagueness of the notion of a characteristic.* We have seen that Salmon's approach to causality is to give an informal characterisation of the concepts of 'production' and 'propagation.' In these characterisations, the primitive notions include 'characteristic,' but nothing precise is said about this notion. While

Salmon is entitled to take this informal approach,[11] in this case more needs to be said about a primitive notion such as 'characteristic,' at least indicating the range of its application, because the vagueness renders the account open to counterexamples.

For example, in the early morning the top (leading) edge of the shadow of the Sydney Opera House has the characteristic of being closer to the Harbour Bridge than to the Opera House. But later in the day (at time t, say), this characteristic changes. This characteristic qualifies as a mark by IV, since it is a change in a characteristic introduced by the local intersection of two processes, namely, the movement of the shadow across the ground, and the (stationary) patch of ground that represents the midpoint between the Opera House and the Harbour Bridge. By III, this mark, which the shadow displays continuously after time t, is transmitted by the process. Thus, by II, the shadow is a causal process. This is similar to Sober's counterexample, where a light spot 'transmits' the characteristic of occurring after a glass filter is bolted in place (1987: 254).

So there are some restrictions that need to be placed on the type of property allowed as a characteristic. Having the property of 'occurring after a certain time,' (1987: 254) or the property of 'being the shadow of a scratched car' (Kitcher 1989: 638) or the property of 'being closer to the Harbour Bridge than to the Opera House' (Dowe 1992c: sec. 2.2) can qualify a shadow to be a causal process. There is a need to specify what kinds of properties can count as the appropriate characteristics for marking. It is not sufficient to say that the mark has to be introduced by a single local interaction, for as the earlier discussion suggests it is always possible to identify a single local interaction.

The difficulty lies in the type of characteristic allowed. A less informal approach to the subject could have provided, for example, a restriction of 'property' to 'nonrelational property,' thereby avoiding this particular problem.[12]

---

11. Hanna (1986: 584) claims that Salmon's informal approach makes his account immune to any counterexample or independent test of adequacy. But, unlike the counterexamples to be presented here, Hanna's counterexamples involve non-actual possible worlds.
12. A possible objection to this counterexample that Paul Humphreys (private communication) has led me to recognise is that, by III, for a mark to be transmitted it must be exhibited in the absence of interactions; yet this is not so: the type of interaction by which the mark was said to be introduced (the intersection of the shadow and the patch of ground) continues to occur after time t. Intuitively, of course, this objection is correct. But it in fact serves to highlight another sense in

There are a number of possible ways to provide a more precise account of 'characteristic,' either in philosophical terms such as 'property,' or in terms of precise scientific notions such as 'molecular structure,' 'energy,' or 'information.' In physics and chemistry description of the *structure* of a molecule, or larger solid body, is given in terms of geometrical arrangement as well as of the constituent particles and bonding forces. In biology the *structure* of a cell refers to its geometry as well as to its constituents. Clearly, a specific definition such as 'chemical structure' is not broad enough for Salmon's purposes. Although he uses examples such as a drug that causes a person to lose consciousness because it retains its 'chemical structure' as it is absorbed into the bloodstream (1984: 155), it is nevertheless clear that the 'structure' of a car, a golf ball, a shadow, or a pulse of light is not simply 'chemical structure.' But perhaps this suggests a general characterisation in terms of constituent material, bonding forces and geometrical shape. I believe such an account has a lot of potential. For example, a chalk mark on a ball is a change in constituent material, a dent in a car is a change in geometrical shape, and so on.

A different approach would be to link 'characteristic' to 'property,' of which there are precise philosophical accounts available (see, for example, Armstrong 1978). Rogers takes this approach, defining the state of a process as the set of properties of the process at a given time (1981: 203). A 'law of noninteractive evolution' gives the probability of the possible states at a later time, conditional on the actual state. This is what it means to characterise structure (1981: 203). However, in *Scientific Explanation and the Causal Structure of the World*, Salmon carefully avoids using the word 'property' (1984: chaps. 5, 6), possibly so as to avoid certain logical or set-theoretical paradoxes (1984: 61, note 7). Salmon has since indicated that application of his approach to the specification of admissible types of properties for determining objective homogeneity would be fruitful here (1985: 652). Salmon has more recently elaborated on how that account should go: the sorts of properties that are permitted are those from an objectively codefined class (1984: 68, definition 2), "which is explicated in terms of physically possible detectors attached to appropriate kinds of computers that receive carefully specified types of information" (1994: 300–301). While I have

which Salmon's theory is too vague: what type of interaction should count in III, and in IV? For in reality no process moves with a complete absence of local causal interactions.

sympathy for the intent here, I'm not sure that the notion of an objectively codefined class really captures the desired distinction. It will rule out some of the counterexamples deployed in my analysis, but only some. 'Being the shadow of a scratched car' is not an objectively codefined class, because that cannot be determined locally, but 'being the shadow of a car with its bonnet up' can be determined by a local physical detector-computer. Properties like shape and speed, which can be introduced locally and 'transmitted,' can be discerned by detector-computers. I suspect that Salmon hasn't ultimately rid Reichenbach's 'codefined class' of its epistemic essence, because such properties have their ontic grounds elsewhere, yet are discernible locally.[13]

However, even if that approach were successful, there are further difficulties of a different kind. First, there are cases of "derivative marks" (Kitcher 1989: 463), where a pseudo process displays a modification in a characteristic on account of a change in the causal processes on which it depends. This change could either be at the source, or in the causal background. A change at the source would include cases where the spotlight spot is 'marked' by a coloured filter at the source (Salmon 1984: 142) or a car's shadow is marked when a passenger's arm holds up a flag (Kitcher 1989: 463).

The clause 'by means of a single *local* interaction' is intended to exclude this type of example: but it is not clear that this works, for does not the shadow intersect with the modified sunlight pattern *locally*? It is true that the 'modified sunlight pattern' originated, or was caused by, the passenger raising his arm with the flag, but the fact that the marking interaction is the result of a chain of causes cannot be held to exclude those interactions, for genuine marking interactions are always the result of a chain of causal processes and interactions (Kitcher 1989: 464). Similarly, there is a local spacetime intersection of the spotlight spot and the red beam.

However, even if the 'local' requirement did exclude these cases, there are other cases where pseudo processes can be marked by changes in the causal background that are local. For example, imagine that there is a long stretch of road where the side of the road is flat, then farther along there is a long fence close to the road. The shadow

13. On the other hand, if the attempt did work, then I think the notion of mark transmission would be superfluous, since only causal processes could possess the right sort of property. As we shall see in the following chapter, this is exactly how my account works.

of the car will change its shape abruptly as the car reaches the fence. We can say that there is a local interaction between the shadow and the beginning of the fence that produces a permanent modification to the shadow. A similar case is where a person runs around an astrodome holding up a red filter that modifies the spot. The clause 'by means of a *single* local interaction' is intended to block these cases: but if this is employed too heavily it would exclude causal processes as well, such as Salmon's paradigm case where a red filter modifies the light beam. In this case the filter continues to act on the beam, just as in the fence or moving filter cases.

In any case, it is possible to modify a pseudo process by a single interaction: take the case where a stationary car (a causal process) throws its shadow on a fence. Suddenly the fence falls over, producing a permanent modification in the shadow. Then the shadow has been marked by the single local action of the falling fence. Salmon's counterfactual requirement that the process remain uniform (presumably in the absence of the marking interaction, all other things being equal) does not help in these cases: the shadow would have remained uniform had the fence not fallen. Indeed, most of the above cases fulfill this counterfactual requirement. Further, these cases are not ruled out by attempts to restrict the kinds of admissible properties by admitting only those that can be detected by physically possible detectors, since the relevant property here is shape, which certainly is detectable by physical detectors. So there does not seem to be any obvious way of answering this difficulty. Thus there are two separate classes of pseudo processes (in Kitcher's terminology, derivative marks and pseudo-marks) that qualify as causal according to the *MT* principle. I conclude that the mark theory fails to adequately capture the distinction between causal and pseudo processes.

IV.3.3 Statistical Characterisation of Causal Concepts

In this section the whole approach of using statistical relations to analyse causal mechanisms will be challenged. It will be argued: (1) that causal production cannot be analysed in terms of statistical relations, and (2) that causal interactions cannot be analysed in terms of statistical relations.

First we present the argument that the conjunctive fork does not adequately characterise causal production. The main argument is that the conjunctive fork is not essential for, nor does it entail, causal production. First, the conjunctive fork is not essential for the production of structure of a causal process. The conjunctive fork characterises two effects arising from a common cause, although it could be generalised to more than two effects (Salmon 1984: 161, note 3). But it has no application where a single effect is concerned, for there is no correlation. Yet surely if only one actor gets food poisoning, or only one student hands in a plagiarised paper, we still have cases of production of structure. The development of bacterial poisoning is a causal process whether there are ten, twenty or only one afflicted. Clearly there are cases of production of structure of causal processes where no conjunctive fork can be recognised. The statistical characterisation is therefore not essential. Second, the existence of a conjunctive fork does not entail the presence of a common cause, and therefore does not entail the presence of a causal process or causal production. Salmon himself argues this (for example, in "Crasnow's Example," 1980: 59.) In fact, neither Reichenbach nor Salmon ever claimed that every conjunctive fork represents a common causal fork, as Torretti appears to suppose (1987). Thus there may be an event that fulfils equations (1)–(2), but that is not a common cause. Therefore the conjunctive fork does not entail the production of structure of causal processes.

Third, we can argue that, as well as being unnecessary and insufficient for causal production, the conjunctive fork is not even normally associated with cases of causal production. Salmon's own examples inadvertently illustrate this:

When we say that the blow of a hammer drives a nail, we mean that the impact produces penetration of the nail into the wood. When we say that a horse pulls a cart, we mean that the force exerted by the horse produces the motion of the cart. When we say that lightning ignites a forest, we mean that the electrical discharge produces a fire. When we say that a person's embarrassment was due to a thoughtless remark, we mean that an inappropriate comment produced psychological discomfort. (1984: 139)

In none of these cases is there any statistical correlation, let alone a conjunctive fork. This is to be expected, of course, if causal production is associated with individual, lawful causal processes. But even when conjunctive forks are associated with cases of production of structure, they do not enlighten us about the production of structure. The conjunctive fork points to a common cause, which explains an amount of 'order' in the world, but it does not explain or explicate the structure of a causal process. For that, one would expect at least a description or governing statement about how a constancy of some property over time (structure) arises. This might involve description of how a process came to have a property, or how a property came to persist, or something similar. The existence of a correlation with an explanatory common cause does not enlighten us at all about such considerations. For example, the conjunctive fork might govern the production of correlations between simultaneous instances of food poisoning, or plagiarism, but tell us nothing about the development of bacterial infection or about the production of an assignment from a fraternity file. It is completely inadequate to this task. The production of order and the production of structure are not the same thing. Conjunctive forks may characterise the former, but they do not characterise the latter.

To labour the point, we might agree that the molecular hypothesis explains the remarkable coincidence that the same value of Avogadro's number can be calculated from experiments on Brownian motion and from experiments on X-ray diffraction (Salmon 1984: 217–218). But this is not to say that the conjunctive fork characterises the production of structure associated with the processes in these experiments – that is, the bombardment of macroscopic particles by the molecules of the medium, or the propagation of X-rays through a grating and onto a screen. In the latter case production of structure would be involved, for example, in the effect the grating has on the X-ray. The conjunctive fork provides an analysis of the correlated feature – the determination of Avogadro's number – but it does not provide an analysis of causal production of structure.

There are other inconsistencies in Salmon's attempt to characterise the production of causal structure by the conjunctive fork. To start with, 'structure' is associated with individual causal processes, whereas the conjunctive fork concerns classes of events, and there are difficulties for Salmon's theory in transferring the probability to the single-case in instances where no suitable reference class is available (Dowe 1990: 51–54).[14] Also, according to Salmon the conjunctive fork is a 'nonlawful' phenomenon, but causal processes are governed by physical laws (Salmon 1984: 179). It is hard to see how something nonlawful can characterise an aspect of lawful phenomena. The conclusion, then, is that the conjunctive fork does not adequately characterise causal

14. For a further criticism see Mellor (1971: 54).

production. We deny that conjunctive forks "play a vital role in the production of structure" (1984: 179).

We now present the case that causal interactions cannot be given a statistical characterisation. Salmon introduced the interactive fork to account for cases of common cause where the effects are more strongly correlated than allowed for in equation (2) (1978: 692–694). Salmon also recognised that the interactive fork could be put to further use, as he puts it, "to fill a serious lacuna in the treatment up to this point" (1978: 694), namely, as a characterisation of causal interactions. I wish to contest the success of the latter program. It should be pointed out that Salmon has now changed his mind about a statistical characterisation of interactions. Initially, Salmon confirmed the view he had presented in 1978 (1980: 67; 1982b: 264–265). However, in ensuing papers we find that it is "in many cases" (1982a: 60) that interactions can be characterised by the statistical relations; and indeed, in *Scientific Explanation and the Causal Structure of the World*, the principle *CI* contains no reference to any statistical relation (1984: 171). Salmon in fact comments, "I now think that the statistical characterisation is inadvisable" (1984: 174, note 12). So Salmon is now in agreement with the conclusion that I am arguing for, and in fact he has recently given his own reasons.[15] Nevertheless, it is worth presenting the full argument, which I believe constitutes a compelling case.

Four reasons will now be presented for accepting the conclusion that causal interactions cannot be characterised by the interactive fork. The first two reasons follow similar reasoning to the case concerning conjunctive forks – that the interactive fork is neither essential for, nor does it entail, a causal interaction. First, many causal interactions do not exhibit statistical correlations. Two men collide in the corridor. One gets a bruise on his arm, the other drops his papers. This is an example of a causal interaction, because we have an intersection of two causal processes (the world lines of the two men), and we have mutual modification. Yet no one would expect to find any useful statistical correlations between bruises and sets of disordered papers. Causal interactions do not always fulfill the interactive fork.

15. Initially in (1990). There he explains the primacy of 'aleatory' over 'statistical' causality and illustrates with a number of examples, and further, deftly construes his arguments against probabilistic causality (see sec. 2.4) as an argument against the idea that interactive forks may be characterised by statistical forks. As Salmon notes (1994: 301), he and I now have no disagreement on this point.

Second, interactive forks do not always indicate the existence of a causal interaction. If the lead actor gets food poisoning (A), and the stagehand gets food poisoning (B), then the following relations holds, where D is the event that both join the tour:

$$P(A.B) > P(A).P(B)$$
$$P(A.B|D) > P(A|D).P(B|D)$$

Then we have an interactive fork among events A, B and D, yet this would not qualify as a causal interaction. D is not a genuine common cause, and in any case the correlation arises from de facto background conditions, not as the result of a causal interaction. Therefore the existence of an interactive fork does not entail the existence of a corresponding causal interaction.

Third, the statistical element is discordant with Salmon's more recent account, where causal processes transmit a propensity to enter into interactions. A causal interaction is essentially an individual event, while the interactive fork concerns general events. It is not surprising, then, that Rogers argues that the probabilities involved in statistical forks (relative frequency) are different from the probabilities in probabilistic laws governing processes (propensities). Salmon's account of probability considerations is discussed later in this chapter, but we must note here that Salmon has now replaced the idea of 'positive statistical relevance' with 'propensity' to characterise indeterminism (1980: 70).

Fourth, we would like to be able to include in any definition of causal interactions types of interactions other than those that produce modifications to two processes. Humphreys points out that many interactions produce modification to only one of the two intersecting processes (1986: 1213), for example, a reflected billiard ball. Unfortunately, this is actually an X-type interaction. A genuine example of a Y-type interaction is the decay of radium-226 to radon:

$$^{226}_{88}\text{Ra} \rightarrow {}^{222}_{86}\text{Rn} + {}^{4}_{2}\text{He}$$

Salmon himself expresses a desire to incorporate λ-type and Y-type interactions (1984: 182). Unfortunately, Salmon's causal interactions are defined in terms of two and only two processes. At this point the statistical account is too restrictive. The preceding four points, then, give us more than sufficient reason to abandon the attempt to characterise causal interactions in terms of the interactive fork.

Salmon often expressed the desire to avoid "hidden powers" (1984: 137, 147, 155; 1994), by which the cause brings about the effect. In using this term, Salmon is alluding to Hume's negative critique of the concept of causation (see Chapter 2) and the inherent constraint on the philosophical account of causation, namely, that it must not appeal to features of which we have no positive contentful conception. Salmon's desire to avoid hidden powers shows that he wishes to provide an account that is Humean, in the sense that it entails that causality supervenes on actual particular matters of fact. But there is doubt as to whether his theory achieves this, at two points: in his use of counterfactuals and in his use of propensity. A number of commentators have raised objections to Salmon's claim to have an account of causality that makes use of counterfactual statements, yet which involves no non-Humean necessary connections (Fetzer 1987: 607; Giere 1988: 446; Kitcher 1989; Sayre 1977: 205–206).

Salmon explicitly formulates the principles *MT* and *CI* in terms of counterfactuals. Propositions I to V omit an element of this counterfactual version, yet still involve modality, because II involves the word 'can,' which entails counterfactual statements. Nevertheless, we now provide the counterfactual versions of propositions III and V to bring these into line with *MT* and *CI*.

IIIC. A mark is transmitted over an interval when it appears at each spacetime point of that interval in the absence of interactions, if the mark would not have appeared had the interaction not occurred.

VC. An interaction is an intersection of two processes where both processes are marked, but would not have been had the interaction not occurred, and those marks are transmitted beyond the locus of the intersection.

Clearly Salmon's theory involves counterfactual claims. How can these be construed in a way that does not appeal to non-Humean necessary connections? Salmon's answer is that the kinds of counterfactuals required are those for which well-designed experiments are able to establish the truth values (1984: 149–150). That is, the truth values of these counterfactuals are established inductively from experiments. There are several possible objections to this answer. Fetzer claims that, without necessary connections, generalisations about causal processes that provide ontic grounds for deciding the truth values of counterfactuals cannot be established inductively (1987: 607–608). Now, while Salmon would not agree that a Humean solution to the problem of counterfactuals is impossible, still, it is true that appeal to experimentation provides only epistemic, not ontic grounds for the truth conditions of counterfactuals. Another difficulty is one raised by Salmon himself. Salmon points to the well-known difficulty that truth conditions for counterfactual conditionals are unavoidably pragmatic (1984: 149). As Salmon is aware, this threatens the claim that causality is an objective matter of fact. His answer, however, seems inadequate: scientists determine which conditions are to be fixed. The result: objective truth conditions for counterfactuals. Unfortunately, Salmon does not elaborate on how an otherwise pragmatic decision becomes objective when made by a scientist. Thus he has not shown adequately how counterfactuals are to be construed in a Humean fashion.

The second area concerns propensities. Here, too, Salmon needs a clear and consistent Humean account to sustain his claim. Salmon allows propensity, not as an interpretation of the probability calculus, but as a real tendency, or disposition, associated with and transmitted by a causal process, for a certain interaction under given circumstances. Thus propensity is, for Salmon,

the strength of the tendency of the connecting (probabilistic) causal process to produce the outcome in question (1979: 213)

or

the strengths of the probabilistic causes that produce the various possible outcomes. (1988a: 31)

Propensities are probabilistic causes, the fully objective, relational properties of causal processes. They are not directly observable, but our scientific theories purport to provide their values. They are *not* probabilities (1979: 212; 1982b: 248; 1984: 204–592; 1988a: 14; despite 1984: 203).

The consistency of Salmon's account of propensity has been questioned by a number of critics. Humphreys (1986: 1214) thinks that Salmon has erred in preferring a frequency interpretation of conjunctive forks, while insisting on a propensity interpretation of interactive forks. Similarly, Fetzer concludes that Salmon's dual interpretation means that he has "completely failed to specify an appropriate characterisation of the causal relevance relations" (1987: 609). Since Salmon devotes several chapters (1984: chaps. 5, 6, 7) to characterising causal relevance relations in terms of processes, interactions, transmission and forks, it is fair to presume that Fetzer means Salmon has failed to specify an appropriate characterisation of the interpretation of probability statements within his characterisation of the causal relevance relations.

These questions have been addressed by Salmon (see 1988a). In his article Salmon makes it clear that there is no 'dual interpretation' of probability intended. Statistical relevance data is to be understood as relative frequencies, that are generated by probabilistic causes (propensities) that are not themselves probabilities. And the probability statements in both conjunctive and interactive forks are to be understood as relative frequencies, yet both these forks describe processes that carry propensities which generate the frequencies. Thus Salmon is not directly open to the charge laid by Humphreys and Fetzer. However, one might ask the question, why cannot propensities be probabilities? And, if they are not probabilities, what are they? They look like probabilities, and they have a close relationship to other probabilities, according to Salmon's theory. For example, propensities generate frequency data (1988a: 25), and personal probabilities are usually estimates of propensities (1988a: 32). Worse still, the 'weight' with respect to an objectively homogeneous reference class is identified as the propensity (1988a: 24). Salmon's reason for refusing to allow propensity as an interpretation of the probability calculus is that the propensity makes no sense of inverse probabilities, an argument due to Humphreys (1985). However, this leaves Salmon with a somewhat confused account.

Salmon has been criticised directly for invoking 'physical modalities' in his theory of propensity. Humphreys writes, "To introduce into the singular case a feature (a probability distribution) which is impossible to construe in actualist terms is . . . to introduce elements of modal realism which are at odds with the robustly scientific orientation which makes the rest of the book so attractive" (1986: 1214). According to Fetzer (1988: 128; and see Armstrong 1983: 8–9), Humphreys's objection seems to be based on a view that Armstrong calls 'actualism,' that a property is only real if it is instantiated, and he argues that this is too extreme: it would rule out all dispositional properties, which surely is undesirable. Salmon himself avoids the term 'disposition,' but there must be a close connection between dispositions and propensities.

Again, the problem concerns Humean violations, if Salmon is to maintain that causal facts supervene on actual facts. Presumably Salmon's answer would be along the same lines as his approach to counterfactuals: any analysis of propensity must take account of the way in which experiment is used within science to test the truth of propensity claims, which may, of course, involve counterfactuals. This leaves us with the same difficulties that were involved with his use of counterfactuals. Therefore it is concluded that Salmon has not demonstrated that his theory avoids non-Humean hidden powers.

IV.4 SUMMARY

It appears that we need a different approach to specifying the kinds of physical properties connected to causality. Given our investigations of the transference theory in the previous chapter, one may be tempted at this point to introduce energy or momentum as the kind of characteristic that is transmitted by causal processes and modified by interactions, especially since Salmon claims that "all and only causal processes transmit energy" (1984: 146).[16] As we shall see, the theory presented in the next chapter partly follows this thought.

This would solve several of the most serious difficulties we have been considering, for the single reason that spots and shadows do not transmit energy. A moving spot, which 'transmits' a mark when someone runs along with a red filter, does not transmit energy, so the requirement that causal processes move in the absence of interactions need not be retained, as causal processes that need interactions to sustain them nevertheless transmit energy. The processes with pseudo properties such as 'being closer to the Harbour Bridge than to the Opera House' will not qualify as causal because they do not transmit energy. Processes – such as the car's shadow modified by a single local interaction when the fence falls down – will not qualify as causal because they too do not transmit energy.

However, in Salmon's view there is a problem in using energy as the distinguishing mark of a causal process, because we cannot tell the difference between a case of energy transmission and cases where energy appears in a regular fashion (1984: 146). He gives the example of a spotlight 'spot' moving along a wall – where energy associated with the spot

16. Indeed, Forge (1985: 457) reads Salmon as saying that 'causal process' could be defined as a process that transmits energy.

appears at all the places occupied by the spot. It would seem that the spot 'transmits' energy, since transmission is to be understood in terms of the 'at-at' theory. This is an objection that will have to be answered, as we turn in the following chapter to an attempt to characterise causal processes in terms of properties such as energy and momentum.

In the previous chapter we considered the twin notions of causality as persistence, such as the kind of causality that is related to having identity through time; and causality as transference, where one body affects another. It was argued that an adequate account of causation should account for both modes, and the transference theory was found to be wanting on the score that it rules out the possibility of causality as persistence. At the start of this chapter, in the discussion of Russell's account of causal lines, it was noted that the idea of a causal line captures what was missing in the transference theory – a notion of immanent causality – but that if causal lines are the whole story on causality, then that account loses as much as it gains, for it now ignores transient causality.

Salmon's account of causation in terms of causal processes and interactions allows nicely for both modes of causality. The concept of a causal process fits perfectly with the kind of causation involved in identity over time, or the immanent causality displayed by the spaceship moving by its own inertia. On the other hand, the concept of a causal interaction accounts well for the idea of transference – since two processes are mutually modified in an interaction. On this score, at least, Salmon's theory appears to be superior to the others discussed. Even if the mark theory itself is inadequate, the approach of treating causality as a characteristic of processes and interactions seems to be the right one.

5

The Conserved Quantity Theory

It is profitable to distinguish three key questions about causation. The first question is *what are causal processes and interactions?* I follow Salmon in the view that it is advantageous to focus on this question rather than on more traditional questions about causation. As we have seen, the key task in addressing this question is to distinguish causal from pseudo processes. In this chapter I offer an account of causal processes and interactions that, I argue, adequately makes this distinction. I show how this account answers a range of objections, in comparison to other theories, in particular to Salmon's theory and his recent revisions.

The second question – *what is the connection between causes and effects?* – is not addressed in the present chapter. In Chapter 7, however, I discuss the kind of answer one can give to this second question if one accepts the results of the present chapter; and I defend that answer against its rivals. The third question is *what distinguishes a cause from its effect?* In Chapter 8 I discuss the kind of answer one can give to this if one accepts the results of the present chapter; and I defend that answer against its rivals. It is important to emphasise that the account of causal processes and interactions given in this chapter is not intended to address the second and third questions.

In this chapter an outline of a theory of causal processes and interactions is presented. The approach to be taken is to modify Salmon's theory by introducing the concept of a *conserved quantity*. The central idea is that it is the possession of a conserved quantity, rather than the ability to transmit a mark, that makes a process a causal process. Insofar as it links causation to quantities like energy and momentum, this account also bears some resemblance to the transference theory.

We begin this section with an outline of this Conserved Quantity (CQ) theory,[1] which will be followed by some comments expanding on the intended meaning of the terms used, and some examples.

The conserved quantity theory can be expressed in just two propositions:

CQ1. A *causal process* is a world line of an object that possesses a conserved quantity.

CQ2. A *causal interaction* is an intersection of world lines that involves exchange of a conserved quantity.

A *process* is the world line of an object, regardless of whether or not that object possesses conserved quantities. A process can be either causal or noncausal (pseudo). A *world line* is the collection of points on a spacetime (Minkowski) diagram that represents the history of an object. This means that processes are represented by elongated regions, or 'worms,' in spacetime. Such processes, or worms in spacetime, will normally be timelike; that is, every point or time slice on its world line lies in the future lightcone of the process's starting point. However, it is at least conceivable that the world line of an object may sometimes appear on a spacetime diagram as a spacelike worm. One example of this is the short-lived string. Imagine that a 1,000-mile-long string extended roughly in a straight line spontaneously comes into existence, but then is annihilated one millisecond later. This short-lived string will be a worm in spacetime, but it is not extended far in time. But that worm is the world line of an object, so it is a process on the present account. Another example of particular relevance is a case of a pseudo process, like the spot moving along the wall, which can travel faster than the speed of light. In the case where it does in fact travel faster than the speed of light the process is represented by a spacelike worm. This also counts as a process. Thus, on the present account a process is a worm in spacetime, be it timelike or spacelike; just provided that worm is the world line of an object.

1. As originally given in Dowe (1992c), but including some slight modifications prompted by Salmon's (1994) analysis. Although Brian Skyrms, in his 1980 book *Causal Necessity* (1980: 111), was the first to suggest a conserved quantity theory, the first detailed conserved quantity theory did not appear until 1992 (Dowe 1992a; 1992c).

An *object* is anything found in the ontology of science (such as particles, waves and fields), or common sense (such as chairs, buildings and people). This will include noncausal objects such as spots and shadows. A process is the object's trajectory through time. That a process is the world line of an object presumes that the various time slices of the process each represent the same object, at different times; thus it is required that the object have identity over time. The requirement of identity over time of an object rules out certain worms in spacetime: not every worm in spacetime counts as a process, for not every worm in spacetime is the world line of an object. One type of worm that does not qualify as a process is a timewise gerrymander – an alleged object defined in different ways at different times (see section 5.3). On the present account a timewise gerrymander is not a process, for it is not the world line *of an object*, since objects must exhibit identity over time. Thus Quine's characterisation of a physical object as an intrinsically determinate portion of the spacetime continuum (Quine 1965: 229–231) will not suffice, since it admits as objects timelike gerrymanders.

Worms in spacetime that are not processes I call, borrowing Kitcher's (1989) terminology, 'spatiotemporal junk.' Thus a line on a spacetime diagram represents either a process or a piece of spatiotemporal junk, and a process is either a causal or a pseudo process. In a sense, what counts as an object is unimportant; any old gerrymandered thing qualifies (except timewise gerrymanders). In the case of a causal process, what matters is whether the object possesses the right type of quantity. A shadow, for example, is an object, but it does not possess the right type of conserved quantities; a shadow cannot possess energy or momentum. It has other properties, such as shape, velocity and position, but possesses no conserved quantities.[2]

A *conserved quantity* is any quantity that is governed by a conservation law, and current scientific theory is our best guide as to what these are. For example, we have good reason to believe that mass-energy, linear momentum, and charge are conserved quantities (see section 5.2).

An *intersection* is simply the overlapping in spacetime of two or more processes. The intersection occurs at the location consisting of all

2. The theory could be formulated in terms of *objects*: there are causal objects and pseudo objects. Causal objects are those that possess conserved quantities, pseudo objects are those that do not. Then a causal process is the world line of a causal object.

the spacetime points that are common to both (or all) processes. An *exchange* occurs when at least one incoming, and at least one outgoing process undergoes a change in the value of the conserved quantity, where 'outgoing' and 'incoming' are delineated on the spacetime diagram by the forward and backward light cones, but are essentially interchangeable. The exchange is governed by the conservation law, which guarantees that it is a genuine causal interaction. It follows that an interaction can be of the form X, Y, λ, or of a more complicated form.

'*Possesses*' is to be understood in the sense of 'instantiates.' An object possessing a conserved quantity is an instance of a particular instantiating of a property. We suppose that an object possesses energy if science attributes that quantity to that body. It does not matter whether that process transmits the quantity or not, nor whether the object keeps a constant amount of the quantity. It must simply be that the quantity may be truly predicated of the object.[3]

As expressed in the two propositions just given, the CQ theory aims to provide an answer to the first question, viz., what are causal processes and interactions? In particular, it aims to distinguish causal from pseudo processes, and it does this by distinguishing objects that possess conserved quantities from those that don't. As in Salmon's theory, causality is treated fundamentally as a property of processes and interactions.

We may also include a broader sense of 'causal process,' where a series of causal processes and interactions form a unified sequence. Sound waves and water waves will qualify as causal processes in this sense.

We now turn to some examples.

Example 1. Consider a transmutation reaction where a nitrogen atom ($^{14}_{7}N$) is hit by an alpha particle ($^{4}_{2}He$), producing an oxygen atom ($^{17}_{8}O$) and a proton ($^{1}_{1}H$). The nuclear equation (with Q representing the extra energy needed for the interaction) is given by

$$^{4}_{2}He + ^{14}_{7}N + Q \rightarrow ^{17}_{8}O + ^{1}_{1}H$$

---

3. In previous formulations (Dowe 1992a: 126; 1992b: 184; 1992c: 210) the word 'manifests' was used in place of 'possesses,' but, as D. M. Armstrong has pointed out, this gave the misleading impression that the quantity had to be experienced by human observers (personal communication). This improvement is also suggested by Salmon (1994).

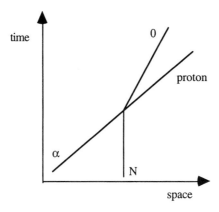

Figure 5.1. Transmutation.

The subscript is the atomic number – the number of protons in the atom, which determines the element type. The atomic number also is the charge of the atom, since the number of protons is equal to the number of electrons. The superscript is the atomic mass, which can vary for an element. The spacetime diagram for this is shown in Figure 5.1.

By definition CQ2, this reaction is a causal interaction, because we have the intersection of world lines where charge, represented by the subscripts, is exchanged. Amongst other things one unit of charge is transferred from the α particle to the nitrogen atom, changing it in the process. So each of the processes involved is a causal process by definition CQ1, because they each possess charge. Note that the N-atom possesses a net charge of zero. On the present view, however, this still counts as possessing a conserved quantity. There is a difference between possessing a zero sum of a quantity and being the sort of object that does not possess conserved quantities. (More on this later.)

Example 2. An example of a Y-type interaction is the decay of radium-226 to radon mentioned earlier, as shown in Figure 5.2:

$$^{226}_{88}\text{Ra} \rightarrow \,^{222}_{86}\text{Rn} + \,^{4}_{2}\text{He}$$

This qualifies as a causal interaction by CQ2 because there is an exchange of charge, where the charge of the incoming process is divided between the two outgoing processes. The three processes involved are all causal by CQ1 because they each possess charge.

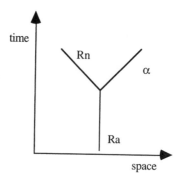

Figure 5.2. Radium-226 decay.

Example 3. As an example of a pseudo process, consider the phase velocity of water ripples. When a stone is dropped into water, the individual waves may travel faster (phase velocity) than the total group of waves (group velocity). Then at the leading edge waves will disappear, while waves will appear at the rear. It is physically possible for phase velocities to travel faster than light, but they cannot be used to convey signals. On our theory these types of phase velocities are not causal processes because they do not possess energy or momentum, or any conserved quantity. The energy, momentum and power of the wave travel at the speed of the group velocity. To generalise, pseudo processes do not possess the type of physical quantities that are governed by conservation laws. Shadows, intersections of rulers and so on do not possess conserved quantities.

V.2 CONSERVED QUANTITIES

A conserved quantity is any quantity that is governed by a conservation law, and current scientific theory is our best guide as to what these are: quantities such as mass-energy, linear momentum, and charge. The idea that the quantities associated with causation are conserved quantities is a suggestion that I present here simply as a plausible conjecture. I have argued in Chapter 3 against Aronson's idea that velocity and certain other physical quantities are the right quantity, and in Chapter 4 against Salmon's idea that an ability to transmit a mark is the right property, but I have no real quarrel with Fair's position that it is energy/momentum. I simply offer the conjecture that other conserved quantities, such as charge, may also serve the function.

94

Conservation laws play the role of identifying which quantities are significant for causation. The claim is not that certain quantities are locally conserved in an interaction or by the process in the absence of interactions, although that will follow. Rather, the account focuses on those quantities that are globally or universally conserved, and connects causality simply to the possession of those quantities.

It is important that conserved quantities be understood in a way that does not appeal to causation, or else circularity threatens. It is common to define conservation in terms of constancy within a closed system. Now if a closed system is simply one with no external causal interactions, that is, a system causally isolated from all others, then we face an immediate circularity. The idea is fine as a rule of thumb – that is, it is true – but it cannot work as an analysis. Instead, we need to explicate the notion of a closed system in terms only of the quantities concerned. For example, energy is conserved in chemical reactions, on the assumption that there is no net flow of energy into or out of the system.

It is important to note that the reference to current theories does not relativise causation to human knowledge – the point is simply that current theories are our best guide to what the conservation laws are. The reason that we cannot simply define a conserved quantity as one that is universally conserved is that some quantity may be accidentally conserved, and such a quantity should not enter into the analysis of causation. Further, regularities are not by any means the only form of evidence about conservation laws – theoretical considerations are also important.

The identity of 'causal process' with 'the world line of an object that possesses a conserved quantity' is contingent, and not metaphysically necessary. The hypothesis is that in our world, and in close enough worlds, such as most of those that obey our laws, a causal process is the world line of an object that possesses a conserved quantity. We leave aside the question of how far we can stray from actuality before this hypothesis stops making sense. In calling this an empirical analysis (see Chapter 1), we emphasise the priority of the claim that the identity holds in actuality. In calling the analysis a contingent identity, we mean that it is contingent on the laws of nature and perhaps even on boundary conditions.

In particular, the theory does not purport to tell us what happens to the identity in distant merely possible worlds. Suppose $\{q_a, q_b, q_c, q_d\}$ is the complete set of conserved quantities in the actual world W_a, and consider a world W_e where none of this set is in fact conserved, and where a conservation law holds instead for q_e. Is the world line of an object in W_e that possesses q_e but none of the set of quantities conserved in W_a a causal process? Or again in W_e, is the world line of an object that possesses q_a, say, but none of the quantities conserved at W_e, a causal process? The answer in both cases is that the theory does not say.

The theory may tell us about closer worlds – for example, those with the same conservation laws as ours. In a world where q_a is conserved, but there is only one object that possesses q_a, the world line of that object is a causal process. Thus the account is not a (Humean) actual-regularity account.

This raises the question of whether the theory is a singularist account (ontologically, not conceptually). I say the account is singularist in the following sense: a particular causal process is not analysed in terms of laws about that type of processes; rather, that a type of process is causal is a matter of generalisation over the particular instantiations of that process-type. The particular is basic.

Thus whether something is a causal process depends only on local facts about the process, namely, the object's possession of a certain kind of physical quantity. It does not depend on what happens elsewhere in the universe, so in that sense being causal is an intrinsic property of a process.

Is this a supervenience (i.e. nonsingularist) account in the sense (e.g., Tooley) that whether the world line of a is a causal process supervenes on whether a possesses a quantity q such that there is a law governing q? No, no such claim has been made. The theory simply says that at this world, just if an object possesses one of the quantities that is actually conserved, then the world line of that object is a causal process. This is a local, particular matter.

Alexander Reuger (1998) has argued that in some general relativistic spacetimes, on the Conserved Quantity theory, it is not a local matter whether a process is causal. Reuger points out that in general relativity, global conservation laws may not hold. In the nonrelativistic case a differential conservation law such as the electrodynamic continuity equation:

$$\text{div } \mathbf{j} = -\partial/\partial t \, \rho$$

(where $\mathbf{j}$ is the current density vector [the amount of electric charge moving through a unit volume in a unit time], such that $\mathbf{j} = \rho \mathbf{v}$, where

v is the charge velocity and ρ is the charge density) entails, via Gauss's theorem, the integral conservation law:

$$\partial/\partial t \int \rho dV = -\int \mathbf{j}\, \mathbf{n}\, dS$$

for a surface S of a volume of integration V. The differential is the local, the integral the global form of the conservation law.

In the general relativistic context, however, a differential conservation law holds for energy-momentum,

$$\nabla^a T_{ab} = 0$$

for the covariant derivative ∇^a, given Einstein's field equations. But unless spacetime possesses special symmetries, there will be no integral formulation. Reuger concludes that whether conservation laws hold is contingent on the global properties of spacetime, and that the choice is therefore either to insist that causation is intrinsic, and that there are no genuine causal processes, or to abandon the intuition that causation is intrinsic to a process or event.

However, there is a third option, which follows from what I have already said. The Conserved Quantity theory is a contingent hypothesis, contingent on the laws of nature, for example. This means if the laws turned out to be a certain way, the theory would be refuted. This may be the case if it turns out that there actually are no conservation laws.

But the fact that there are general relativistic spacetimes in which global conservation laws do not hold does not entail that global conservation laws fail in our world. Whether they do or not depends on the *actual* structure of spacetime, and in particular whether certain symmetries hold. As I understand it, our spacetime does exhibit the right symmetry; global conservation laws do hold in our universe as far as we know. I take it, then, that the conserved quantity theory is not refuted.

I have suggested that the account should probably hold in all physically possible worlds, that is, in all worlds that have the same laws of nature as ours. Has Reuger shown that this is not so? Not at all. To say, for example, that nonsymmetric spacetimes are possible can be misleading. It means simply that it is a solution to the equations of the General Theory of Relativity. But this doesn't mean that such a world is a physically possible world in the sense given here. If such a world violates other laws that hold in the actual world, then that world is not physically possible. This is exactly what we have in these nonsymmetric spacetimes. Symmetries and conservation laws that hold in the actual world break down, so it is not a physically possible world in my sense.

Therefore we need not give up on the Conserved Quantity theory, understood as a contingent hypothesis, nor on the idea that causation is actually intrinsic.

V.3 POSSESSION, TRANSMISSION AND GERRYMANDERED AGGREGATES

In his criticism of the Conserved Quantity theory just presented, Salmon (1994: 308) offers an argument (see also 1984: 145–146) for requiring 'transmits' rather than just 'possesses': Consider a rotating spotlight spot moving around the wall of a large building. This is a classic case of a pseudo process: in theory such a spot could move faster than the speed of light. But the spot manifests energy at each point along its trajectory. Therefore, Salmon's argument goes, we need more than just the regular appearance of energy to characterise causal processes; we need the notion of transmission. In this section I show how the CQ theory avoids this problem without appealing to any notion of transmission.

A spot or moving patch of illumination does not possess conserved quantities. A moving spot has other properties: speed, size, shape and so on; but not *conserved* quantities such as energy or momentum. What possesses the energy that is 'regularly appearing' is not the spot but a series of different patches of the wall. The spot and the patch of wall are *not* the same object. The patch of wall does not move. It *does* possess conserved quantities, *its* world line does constitute a causal process, and *it* is not capable of moving faster than the speed of light. The spot does move, but does *not* possess energy and *is* capable of moving faster than the speed of light. Therefore 'whether or not an object possesses a conserved quantity' is an adequate criterion for distinguishing causal from pseudo processes.

Hitchcock (1995) provides another example of the same objection, where a shadow moves across a charged plate, at every stage manifesting a conserved quantity, charge. The answer to this is the same as for the spot of light.

Salmon (1994: 308) gives an ingenious counterexample to this answer, asking us to consider "the world line of the part of the wall surface that is absorbing energy as a result of being illuminated" (1994:

308). This "gerrymandered" object is the aggregate of all the patches of wall that are sequentially illuminated, taken only for the time that they are being illuminated. Salmon argues that this object does possess energy over the relevant interval, but does not *transmit* energy. The implication is that the world line of this object is not a causal process, yet the object possesses energy; therefore we need to invoke the notion of transmission – possession is not enough.

I think that the objection is misdirected, as I shall now argue. Such gerrymandered aggregates do not qualify as causal processes according to the CQ theory, providing what counts as an object is adequately explicated.

According to the CQ theory, there are causal processes such as billiard balls rolling across tables, and pseudo processes such as shadows and spots of light. Is this exhaustive of all items that may be represented as occupying a spacetime region? The answer is no; there is also "spatiotemporal junk" – items that are not processes at all on the CQ definition. An example is what I earlier called "timewise gerrymanders." A timewise gerrymander is a putative object defined over a time interval where the definition changes over time (the putative object is really different objects at different times). A comparison may be drawn to Goodman's gerrymandered properties (grue, bleen) which are really different properties at different times (Goodman 1955). An example of a timewise gerrymander is the putative object x defined as:

for $t_1 \leq t < t_2$; x is the coin in my pocket
for $t_2 \leq t < t_3$; x is the red pen on my desk
for $t_3 \leq t < t_4$; x is my watch.

Notice that x occupies a determinate spacetime region, and that at any time in the interval t_1 to t_4, x 'possesses' conserved quantities such as momentum (although not strictly speaking, for something must be an object in order to possess a conserved quantity). Clearly, there are innumerable such timewise gerrymanders.

Timewise gerrymanders are to be distinguished from spacewise gerrymanders. An example of a spacewise gerrymander is the putative object y consisting of my watch plus the red pen on my desk plus the coin in my pocket. The spacetime representation of x consists of three vertical lines that do not coexist at any time, whereas the spacetime representation of y consists of three vertical lines that coexist over the entire interval.

Timewise gerrymanders are sometimes defined just by a single formula. For example, in a billiards game, take x to be 'the closest ball to the black ball,' in the sense that, say;

for $t_1 \leq t < t_2$;	x is the pink ball
for $t_2 \leq t < t_3$;	x is the red ball
for $t_3 \leq t < t_4$;	x is the white ball.

Here x is a timewise gerrymander, occupying a spacetime region and 'possessing' conserved quantities. A similar case would be 'the president of the USSR' taken to refer to a single object consisting of the mereological sum of each of the presidents from Lenin to Gorbachev, each taken only for the time they were in office. Other single-formula timewise gerrymanders include 'the object currently in the centre of my field of vision' or 'the object nearest to my car,' provided these are taken in the gerrymandering sense and not in the usual sense as referring not to objects of which the description may be true at times other than the present, but to objects of which the description is currently, if not always, true.

Timewise gerrymanders sometimes display consistency of some feature. For example, in a box of molecules, take x to be whatever molecule has momentum p_x, taken just for the time that it has that momentum. Again, x is a timewise gerrymander, occupying a spacetime region and 'possessing' conserved quantities, but here x also has a *stable* momentum. The object has a consistency of some property over its entire history. Finally, timewise gerrymanders sometimes display spatiotemporal continuity. For example, consider a line of ten contiguous stationary billiard balls. Let x be the mereological sum of the first ball during the first time interval, plus the second during an immediately subsequent time interval, and so on for the ten balls. Then x is a timewise gerrymander, represented on a spacetime diagram by a diagonal line roughly one billiard ball wide. Note that x is not to be confused with the object y consisting of the whole line of ten balls, which *is* a genuine object, and which is represented by a vertical block ten balls wide.

Since the CQ theory has it that the world line of an object possessing a conserved quantity qualifies as a causal process, doesn't x qualify as a causal process? The answer is no, because x does not qualify as an object. As we have seen in section 5.1, there is implicit in the CQ theory a restriction on what counts as an object – it must display identity over time. A timewise gerrymander is a collection of different objects at different times. So, timewise gerrymanders are not objects.

So, turning to Salmon's example of the aggregate of the patches of wall sequentially illuminated, we can see that although it is generated by a single description, involves some uniformity, and displays spatiotemporal continuity, it nevertheless is a timewise gerrymander, and not in fact an object on my definition, since it does not display identity over time. The spot itself is an object (although not causal), and the entire patch of wall is an object (like the ten billiard balls), but the timewise gerrymander is not. It therefore is not a process of any sort, let alone a causal process; it qualifies on my account as spatiotemporal junk.

We need to be careful of an equivocation. The world line of 'the patch illuminated at time t_1' is a genuine process – it is an object that is temporally illuminated. However, the object made up of 'the patch of wall illuminated at time t_1' plus 'the patch of wall illuminated at time t_2' plus . . . is a timewise gerrymander.

Therefore we may conclude that Salmon's example, being a timewise gerrymander, does not prove that the conserved quantity definition is too inclusive, since such timewise gerrymanders do not qualify as objects. Therefore, the example does not force us to supplement the notion of possession of the relevant quantities. Possession may be only nine tenths of the law, but it is the full story on causal processes.

Max Kistler (1998: 16–17) argues that this account is circular in the sense that the requirement that an object display identity through time rules out timewise gerrymanders only if you already know that their temporal parts are not parts of the same object. This is true, but it misses the point of my analysis. I take it that it is intuitively clear that the temporal stages of certain timewise gerrymanders are not temporal parts of a single object. Once that is recognised, it becomes clear that Salmon's series of spots is also a timewise gerrymander. It's up to an account of identity through time to explain why the temporal stages of a timewise gerrymander are not parts of a single genuine object. We discuss such accounts in the next section, although it is not the burden of this book to give a theory of identity.

V.4 IDENTITY THROUGH TIME

The notion of a process, as explicated in the CQ theory, involves the idea of identity through time. A process is the world line of an object, so fundamentally, to constitute a process, an object must persist over time. This analysis presupposes a notion of identity through time; since a world line is the line traced out by an object through time, it is necessary that the object be the same object at different times.[4] If identity appears as a requirement in the theory, then we need to say more about how this concept might be analysed.

Indeed, in rejecting the identity assumption implicit in the conserved quantity theory Salmon says,

I have offered a concept of causal transmission analysed in terms of the "at-at" theory for which Dowe has traded an unanalysed concept of genidentity. This is not, I think, an advantageous exchange. (1997: 468)

Salmon comments on his own revised version (see section 5.6) that "it yields a criterion that is impeccably empirical, and thus it provides an acceptable answer to the fundamental problem Hume raised about causality" (1997: 468).

So Salmon's objection to taking identity over time as primitive in a theory of causal processes appears to be that it violates the empiricist's stricture that one should not invoke empirically inaccessible elements as unanalysed or primitive in a philosophical theory. In this section we consider the main accounts of identity through time, and then return to Salmon's objection.

V.4.1 Strict Identity

One way to analyse identity is as literally strict identity. An object is identical with its other temporal parts in the same way that it is identical with itself, and in the same way that different things have the same property (see Armstrong 1980). According to this view *an object must be wholly present at a time* in order to exist at that time. That is, when you have an object at a time, it's not that strictly speaking you just have a part of that object – a temporal part. If you have an object then you have the whole object at that time.

A major difficulty with strict identity is the problem of 'temporary intrinsics,' in Lewis's phrase (1986: 202–420). If an individual can have contradictory intrinsic properties at different times, how can it exhibit strict identity? One reply is to say that such properties are time-

---

4. Genidentity of objects is also presupposed in the transference theory of Aronson and Fair, because that theory requires that the recipient object is not the same object as the donor object, otherwise there would be no transference. So the objects involved need to be identified through time, as well as the energy/ momentum.

indexed, and that there is no contradiction in having one intrinsic property at one time and another at a different time. One way to explicate this, mentioned (but not endorsed) by David Lewis, is to treat these properties as disguised relations – relations to times.

Salmon's objection would be circumvented simply by adopting the strict identity position. Contrary to a widely held opinion, it doesn't require any unanalysed concepts – for a start, nothing is clearer than self-identity, and in any case, the requirement of identity over time can be stated without using the concept of identity by using a property of totality (I thank David Lewis for this suggestion): there is an object and the whole of it is located at t_2. (See Lewis 1986: 192–193.) So one option is simply to take identity as strict identity (see Dowe 1999).

The alternative to taking objects as wholly present at a time is to take objects as essentially four-dimensional, existing in time in exactly the same way as they exist in space. On this view time slices of the world line are parts of the object – temporal parts. To adopt the four-dimensional conception would require some further way of identifying causal processes as genuine, some way of ruling out timewise gerrymanders. One example is the attempt by Fair to achieve this by appeal to the identity over time of the quantity itself; unfortunately, that notion can be shown to be incoherent.[5] Another is Salmon's appeal to the notion of transmission, as we will see (section 5.5). To do so via the concept of identity, without appealing to strict identity, one must say what is the relation between the temporal parts. Common ways to do this are the similarity-continuity theory and the causal theory of identity. These options will be considered in the next two subsections.

V.4.2 The Similarity-Continuity Theory of Identity

The 'similarity-continuity' approach to identity through time is sometimes called a Humean conception – even though strictly speaking Hume's own theory of identity was a causal theory (see the following subsection) – and it makes particular sense on a Humean metaphysics: viz., that the world is a world of 'bits,' particular local matters of fact, with no logical connections between spatiotemporally separated bits. Since there are no logical connections between the bits, there can be no relations of strict identity across time. However, there can be sur-

5. See Chapter 3, section 3.4.

rogate relations between the bits that can link bits as being temporal parts of the same object. The similarity-continuity account of identity says that bits at different times are connected as temporal parts of the same object, in a looser sense of 'same,' just if appropriate spatiotemporal and resemblance relations hold between the bits. For example, a round black hard object at x at t_1 is the same object as one at y at t_2 if, roughly, the second object is also black, hard and round, and at every spatiotemporal point between x and y there is a hard round black object. Obviously more complicated relations are expected to obtain for more general cases involving more radical changes.

A common line of argument against such Humean accounts is well represented by an argument due to David Armstrong. Armstrong gives an example of two gods who are creating and destroying things independently. By coincidence, one destroys a certain item just at the same instant that the other creates an exact replica of that item at the same location (1980: 76).[6] It may seem to everyone that the same object still exists, but they would all be wrong. It is really a different thing.[7] Armstrong concludes that similarity-continuity theories cannot account for identity through time.

However, this kind of objection does not hold in this context, since we are seeking an empirical analysis. We want to know what causation is in this world, and perhaps in worlds with the same laws of nature as ours, and so what would happen in far distant worlds is of no relevance. In this context there is no need to satisfy criteria for conceptual analysis.

Salmon's objection is met on the similarity-continuity account. There is nothing in this account that can offend the empiricist demand that one should not invoke empirically inaccessible elements as unanalysed or primitive in a philosophical theory.

V.4.3 The Causal Theory of Identity

According to the causal theory of identity (for example, Armstrong 1980),[8] the relation of identity through time involves the relation of

---

6. In that version the 'items' were Richard Taylor and his twin.
7. Of course, if we are attempting an account of identity *in this world* then such counterexamples would not be relevant.
8. A causal theory of identity was defended by Hume, who thought identity through time could be analysed in terms of resemblance, contiguity and causation (*Enquiry*: 246). For Hume, causation reduces to contiguity and resemblance; and resemblance is a relation of ideas.

causation: for an object to display identity over time it is required as a necessary condition that its temporal parts be related as cause and effect.

However, such an account is not available for present purposes. If the CQ theory of causal processes is correct, then the relevant notion of causation is itself dependent on the notion of identity through time, so the causal account of identity is excluded, or else the whole account would be circular. So I reject the causal account of identity through time. I do agree that there is a strong connection between causal connection and genidentity, but I claim that identity is the more basic notion: causal processes cannot be understood except in terms of identity over time.

The causal theory of identity through time has become quite popular in recent years,[9] although it has its critics.[10] Some opponents reject it because it appeals to what they see as the unduly metaphysical notion of causation. But I think that there are a number of more specific considerations that tell against it, and in this section I will present one of those considerations. My conclusion is that in the end it is no great loss that such a theory must be ruled out.

The argument against the causal theory of identity appeals to the idea of a pseudo process. In particular, it turns out that a process can have identity over time without causation. In fact, all (and only) pseudo processes are like this. Consider a spotlight spot moving across a wall. Its temporal parts are not causally related – it is not a causal process. But its temporal parts are related by the identity relation – the spot at time t_1 is the *same* spot that exists at a later time t_2. If there were two spots moving around in a chaotic fashion we might start to wonder which is which, presupposing identity through time. To give another example, my shadow is always my shadow – it has identity through time, but it is not a causal process. Therefore you can have identity without causation, although you cannot have causation without identity. So it is appropriate that our theory of causation rules out the causal theory of identity.

There are two replies available to the defender of the causal theory, but neither is entirely satisfactory. The first reply is to deny that pseudo

---

9. For example, Armstrong(1980), Lewis (1983: chap. 5), Nozick (1981), Parfit (1984), Shoemaker (1984: chap. 11) and Swoyer (1984). "Virtually all leading contributors to the current literature on personal identity . . . subscribe to it," according to Kolak (1987: 339).
10. For example, Ehring (1991), Hirsch (1982) and Kolak and Martin (1987).

processes display identity through time. If we speak and think of the spot or shadow as being the 'same' spot or shadow, then that can only be a manner of speaking that is not grounded in reality. For example, it may be argued that shadows and spots are not genuine things at all, and so they don't endure through time.

But this eliminativism runs into problems with scientific cases of pseudo processes. Here I am thinking in particular of the example, given earlier, of the water ripples moving at the so-called phase velocity. Such ripples can move faster than the speed of light, but do not possess conserved quantities, so they are pseudo processes. But they figure in the scientific description of the world, so they must be taken as genuine processes. So, quite apart from its intuitive awkwardness, this eliminativism faces quite serious difficulties.

The second reply available to the proponent of the causal theory is to distinguish grades of identity in the following way. Top-grade identity always involves causation. But there is also a lower grade of identity, such as that displayed by pseudo processes, which does not directly involve causality. Then we can say that spots and shadows have identity in a sense, although not in the same sense as do billiard balls and persons. So there are two types of identity – the causal and the noncasaul.

This may at first appear completely ad hoc, but in fact it can be given a credible basis. A pseudo process is one that depends in an ontological sense on a causal process or on causal processes. The shadow is the shadow of the car, and the movement of the shadow is dependent on the movement of the car. In fact, characteristics of pseudo processes depend on characteristics of causal processes, in the sense that there can be no difference in the characteristics of a pseudo process without there being some difference in the characteristics of some causal process. So low-grade identity can be understood in this way: an object has low-grade identity over time only if it is dependent in the appropriate way on an object (or objects) that has high-grade identity over time. In other words, a process can have identity over time without causation only if it depends in the appropriate way on a process that does involve causation.[11]

However, while this reply is not entirely ad hoc, it does introduce further complication, and as such represents a loss of theoretical sim-

11. My formulation of this response was spurred by some comments by Michael Tooley (personal communication) on my first argument.

plicity and unity. On this view there are two kinds of identity, whereas before we thought that there was only one. But the proponent of the causal theory has a response to this objection too. For, while the causal theory leads to a loss of theoretical simplicity and unity, it supplies a theoretical unification at another point, which at least compensates for the loss here. The unification that the causal theory supplies is the unification of immanent and transient causality, for on the causal theory one can say that causation as persistence turns out to be a species of transient causation, in the sense that there is a 'transfer' of energy from an earlier time slice of the process to a later time slice, where these time slices are taken as distinct particulars.[12]

So we can see that there are independent reasons for rejecting the causal theory of identity, in addition to the fact that its falsity is entailed by the conserved quantity theory.

V.4.4 Reply to Salmon

We have seen that of the main current approaches to identity over time the most popular, the causal theory, is unsatisfactory; but that either the strict theory or the continuity-similarity theory would do the job, although both have their shortcomings. For this reason I would prefer to leave the notion of identity of an object as primitive in the theory.

But in that case we must face Salmon's objection to the Conserved Quantity theory, that it violates the empiricist requirement that one should not invoke unanalysed or primitive elements. But there are three points to be made in reply to this objection.

First, I admit that there is a need to explain what identity is. I don't want to be committed to the view that identity is an unanalysable primitive. Identity is a placeholder, not a primitive in a deep sense.

Second, I therefore take the position that the question of identity remains open. But philosophy is an ongoing enterprise in which advances always raise new questions. The task of explaining or analysing does not have to reach a terminus point in order for there to be progress. In this respect, philosophy is like science, which, as Musgrave once pointed out (1977), continually provides explanations, but never final explanations.

Third, and more directly, Salmon's position may be open to an (ad

---

12. This idea was suggested by David Armstrong in response to the transference theory (in conversation).

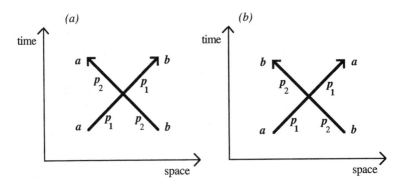

Figure 5.3. (a) A causal interaction. (b) A mere intersection.

hominem) symmetric complaint, namely, can the concept of a 'causal process' be explained without reference to identity? If it can't, then the notion of a causal process is an unanalysed primitive in the theory, and by Salmon's lights that is unacceptable. Thus one question I want to focus on is, can causal processes be explained without reference to identity? If the answer is no, then things are even; if I don't explain identity, then equally the alternative doesn't explain causal processes. If this were the case, then Salmon's charge would be deflected, although we would all agree that there would be a need for further work. I will argue in section 5.6 that this is indeed the case. In the meantime, it is worth thinking briefly about the role that identity plays in science.

Even in our world, identity may or may not be a matter of just similarity or continuity. Nevertheless, scientists do manage to track things accurately through time. In fact, it is a necessary prerequisite for the application of many of the laws of nature that things be reidentified through time. Tracing individuals through time is an integral part of scientific theorising. No account of how this is possible will be given here, but it's important to note that scientists for the most part do perform the task successfully.

Consider an interaction between two billiard balls, as shown in Figure 5.3 (a). The causal interaction involves a collision between two billiard balls, where there is an exchange of momentum. The 'mere intersection' is a case where the balls pass through each other without being affected, a *highly* unlikely occurrence in our world. Let us suppose the two balls are identical, that is, indistinguishable, even to a quite sophisticated level of property detection.

108

According to the CQ theory there is a real difference between (a) and (b) – where the particular balls end up is quite different. Scientific description will distinguish between the two cases. The inference that a given case is really an example of (a), not (b), will usually be made from plausibility considerations based on other physical conditions, not from strict observation,[13] nor by application of the conservation laws.[14] But fundamental to the reasoning is the notion of identity through time.

In Chapter 2 we considered the approach known as the Humean regularity theory, in which causes supervene on regularities or laws. But there's another supervenience thesis that the CQ approach may yet share with the Humean approach. This thesis is that the relevant notion ('cause') supervenes on particular matters of fact.[15] The CQ account may be Humean in this sense, in that causation is just about discernible patterns in the world that amount to objects possessing certain properties. The open question is whether the notion of identity through time, which is primitive in the present theory, can be characterised in a Humean way in terms of particular matters of fact. If identity is understood in a Humean way, then causation is Humean. But if identity is understood in a non-Humean way, then causation is non-Humean. For the present account it is sufficient to note the role identity plays in scientific reasoning.

V.5 ADVANTAGES OF THE CQ THEORY OVER THE TRANSFERENCE AND THE MARK TRANSMISSION THEORIES

In this section we compare the Conserved Quantity theory with the transference theory discussed in Chapter 3, and Salmon's mark theory discussed in Chapter 4. I shall show that the Conserved Quantity theory has a number of decisive advantages over each of these close cousins.

13. Although, in this example, such an observation could be made. A physical detector could in principle be used to detect whether nuclei are passing through or colliding. But a case could be envisaged where single protons, or some sort of fundamental particles, take the place of the billiard balls. Then (b) might be physically impossible, nevertheless the inference in question could perhaps be made on general theoretical considerations, but not on the grounds of observation.
14. Both the examples in Figure 5.3. obey the conservation laws. Of course, we could envisage similar cases where a conservation law is broken; then the conservation laws could be invoked in drawing the inference in question.
15. See Lewis's (1986) characterisation in his introduction.

First, it can be shown that the Conserved Quantity theory avoids the difficulties that beset the transference theory.[16] However, it is important to recognise that there are close connections between the two accounts, and in one sense the CQ theory is a simplification of the transference account. Indeed, the CQ theory salvages the essential truth contained in the transference theory, while remaining noncommittal on some of the issues that cause the problems.

The CQ theory, as it stands, is a minimal account. It characterises the notions of causal process and causal interaction. But even if these are the core, underlying notions of causation, that is not to say that these bare notions provide a full account of every aspect of causation. They do not. The intention is rather to provide a basis upon which it may be possible to build further, in order to account for other aspects of causation, as we shall see in Chapters 7 and 8. (Of course, I do not wish to decide a priori whether there is an objective basis for every aspect of our folk concept.)

First, the CQ theory involves causal interactions, where causal processes intersect, and exchange a conserved quantity. This means that the following relation obtains between the two theories: whenever there is transference in Fair's sense, and (according to Fair) therefore causation, there will be an exchange of a conserved quantity, and therefore a causal interaction. So whenever Fair's theory says there is causation, the CQ theory says there is a causal interaction. The converse does not hold, at least not in all logically possible worlds, because there may be an exchange of some conserved quantity besides energy or momentum, not associated with an exchange of energy or momentum, although I can't think of such a case in worlds that obey our laws of nature.

However, the CQ theory is noncommittal on the question of transfer. It does not require that the quantity be transferred. The notion of exchange is weaker than that; it means merely that there are corresponding changes in the values of the physical quantities. The theory does not say that the object that gave up the quantity is the cause, and the object that gained the quantity is the effect. Thus it is also noncommittal on the question of causation's direction. The CQ theory, as it stands, is symmetric with respect to time, so it does not explain causal asymmetry. It provides a notion of causal connection, but does not say

16. See Chapter 3, section 3.4.

what is the direction of cause to effect. But since the transference theory explains causal direction in terms of the direction of time (Fair) or the empty words 'direction of transference' (Aronson), it is well to be non-committal. Fair's success could be replicated simply by stipulating that the cause is the incoming (earlier) processes, and the effect is the out-going (later) processes. (An examination of ways to extend the CQ theory so as to account for causal asymmetry must await Chapter 8.)

The CQ theory is also noncommittal on the question of the gen-identity of physical quantities. We saw in section 3.4 that this notion is apparently incoherent. The CQ theory simply requires that there is an exchange of some conserved quantity, where exchange is to be under-stood in terms of a change in the value of the quantity. As we have seen, this idea is unproblematic. The difficulty arises if we wish to say that the effect receives the *same* quantity as the cause gave up. The CQ theory is noncommittal on this, and so is consistent with both the view of Aronson and Fair, that physical quantities have identity through time, and the view of Dieks, that physical quantities do not have that type of identity.

Second, the CQ theory involves causal processes, which are the world lines of objects that possess conserved quantities. Following the lead of Salmon, the CQ theory treats as fundamental the notion of a causal process that constitutes the causal structure of the world. This means that the CQ theory, like Salmon's account, provides the basis for understanding that other type of causation, persistence. In section 3.4 we considered an example of a spaceship moving at constant veloc-ity, beyond the reach of any forces. The ship's inertia is the cause of its continuing motion. This kind of causation, ruled out by the trans-ference theory, fits very nicely with the idea of a causal process as the world line of an object that possesses a conserved quantity. In this example, the spaceship possesses momentum. I take it that this is a sub-stantial advantage of the process theories over the transference theo-ries.[17] Like Salmon's theory, but unlike Aronson and Fair's theory, the CQ theory neatly accounts for the Spinozean Disjunction.

17. It is also an advantage over the 'force' theories of causation of Bigelow, Ellis and Pargetter (1988), Bigelow and Pargetter (1990a; 1990b) and Heathcote (1989). In fact, we may turn an argument of Bigelow and Pargetter on its head: they argue that force theories are superior to energy/momentum theories because the latter involve a disjunction. But to account for persistence, the force approach will need to be supplemented, which introduces an ad hoc disjunction; whereas the CQ theory achieves a type of unification in the way it accounts for both sorts of causation.

It can also be shown that the conserved quantity theory avoids the difficulties that Salmon's mark transmission theory faces, in particular those discussed in section 4.4. There we saw that Salmon's theory has problems connected with circularity, with the extensional adequacy of the mark theory, with the statistical characterisation of causal forks, and with non-Humean hidden powers. The conserved quantity theory overcomes each of these difficulties.

First, Salmon's account had problems with circular definitions of 'causal process' and 'interaction.' The conserved quantity theory appeals to 'world line,' 'object,' 'possession' 'onserved quantity,' 'intersection' and 'exchange.' None of these is defined in terms of causal processes or interactions. Hence the theory avoids that kind of problem, provided the notion of identity is not understood in terms of causation. (I have already argued against that approach.)

Second, Salmon's theory of mark transmission was shown to be extensionally inadequate on a number of counts. One problem was that the theory requires that marks be transmitted over a spacetime interval in the absence of further interactions, but in real situations causal processes simply do not evolve in the absence of further interactions. But the conserved quantity theory requires merely that an object possess a conserved quantity, not that it transmit it over any interval or, consequently, that it do so in the absence of further interactions. In fact, a causal process, being the world line of an object of a certain kind, can exist over an extended spatiotemporal interval even if it is involved in causal interactions during that interval, providing the object is still the same object.

Another problem we encountered is that the notion of a mark as an alteration of a characteristic is too vague, and in fact, too inclusive. For example, the shadow of the Opera House is marked at the point where it becomes closer to the Opera House than to the Harbour Bridge. Another example is the shadow of the car being marked when the fence it is projected onto falls over. In each case the motion of the shadow counts as a causal process. The conserved quantity theory does not face this kind of problem, because the relevant properties are restricted to conserved quantities such as energy and momentum. Shadows, for example, do not possess such quantities, although they do have other properties such as speed and shape. Part of the advantage of using 'conserved quantity' rather than Salmon's vague 'structure' clearly lies in its precision: scientific theories give it an exact meaning.

Third, Salmon's attempt to characterise causal forks in terms of statistical structures, such as the conjunctive fork and the interactive fork, runs into a number of difficulties. The conserved quantity theory avoids this kind of difficulty altogether, since it does not appeal to statistical forks at all.

And finally, it was shown that Salmon fails to achieve his goal of providing a Humean account that avoids hidden powers. There were two areas of difficulty; first Salmon's counterfactual formulation results in a non-actualist account, and second, his appeal to propensity fails in the same way. Salmon himself has recently focussed on the counterfactual problem as the most significant difficulty with his mark transmission theory. Salmon writes:

When the mark criterion was clearly in trouble because of counterfactual involvement, it should have been obvious that the mark method ought to be regarded only as a useful experimental method for tracing or identifying causal processes (e.g., the use of radioactive tracers) but that it should not be used to explicate the very concept of a causal process. (Salmon 1994: 303)

The conserved quantity theory does not make use of any counterfactual formulation, and so avoids this difficulty. It does, as mentioned earlier, appeal to the notion of identity over time, and so it would be necessary to give a Humean account of identity before one could finally claim that the conserved quantity theory is Humean in this sense. Nevertheless, it does avoid the immediate problem of counterfactuals.

Neither does the CQ theory appeal to propensity. But I do not deny that causation may be probabilistic. This is a feature that needs development. Whether that development will be Humean depends, of course, on how it is done. I expect that the probabilistic element in the theory must enter as a propensity. An exact analysis of these propensities is beyond the scope of this book: there are, of course, many accounts of propensity available in the literature; see Mellor (1971, 1995) and especially Humphreys (1989: sec. 22). However, contra Salmon and as indicated above, I take it that propensities should be regarded as referring to the operation of objective, indeterministic causal processes and interactions, such that the propensity takes values between 0 and 1 only where there is genuine indeterminism. These propensities would supervene on indeterministic facts about the world, but it remains to be shown how or whether they avoid hidden powers.

There are three further advantages of the Conserved Quantity theory over Salmon's mark theory that are worth mentioning. (1) The theory is also much simpler and less ad hoc. (2) The present definition covers more configurations than did Salmon's definition. On Salmon's account, a causal interaction involves modification to both outgoing processes, an X-type interaction. It is an advantage of the present approach that it includes Y-type and λ-type interactions. (3) The conserved quantity theory does not invoke continuity. Suppose the process involving the decay of one atom to another is not spatiotemporally continuous (perhaps quantum mechanics leads us to think that there are such processes [Forge 1982]). On this point the CQ theory has an advantage over the theory of Salmon since it does not require that processes be spatiotemporally continuous (Dowe 1992c: 214).

V.6 SALMON'S REVISED THEORIES

In a series of papers (1994; 1997), Salmon has replied to the criticisms of his theory (including some of those discussed in the previous section) and offered a further revision that, he argues, is not open to those criticisms (see also Carrier 1998).

The key change concerns the characterisation of causal processes, where Salmon has traded 'the capacity for mark transmission' for 'the transmission of an invariant or conserved quantity.' With characteristic generosity Salmon says that in doing so he is "relying heavily" on my Conserved Quantity theory (1994: 298).

At this point Salmon and I already agree on much. However, in this section I wish to clarify some remaining points of disagreement and offer a defence of my position. In doing so, I raise important questions about the direction of causation and the distinction between causal processes and gerrymandered aggregates. These are questions that have significance for any process theory of causality.

In this section, I summarise the first suggestion of Salmon, then I turn to a defence of the CQ theory. I focus on Salmon's key notion of 'transmission,' arguing that its apparent directionality is a misleading illusion; and I discuss miscellaneous points of difference, including Salmon's requirement that a causal process be spatiotemporally continuous, and his suggestion that invariant rather than conserved quantities best capture the essence of a causal process.

I then discuss Salmon's second revision, and respond to Salmon's main objection to my Conserved Quantity theory – viz., that it appeals to identity through time – by showing that his account also needs that assumption.

V.6.1 The Invariant Quantity Theory

We begin with Salmon's first revised theory (1994; 1998). While aiming to avoid various objections to the mark theory, Salmon nevertheless draws heavily on that earlier account, as well as on the CQ theory. This new account, which I will call the Invariant Quantity theory (or the IQ theory) can be summarised for our purposes (from Salmon 1994: 303, 308) as:

IQ1. A *causal interaction* is an intersection of world lines that involves exchange of an invariant quantity.

IQ2. A *causal process* is a world line of an object that transmits a nonzero amount of an invariant quantity at each moment of its history.

IQ3. A process *transmits* an invariant quantity from A to B if and only if it possesses this quantity at every stage between A and B, in the absence of interactions involving that quantity.

The 'only if' clause in IQ3 was not explicitly given by Salmon, but is needed so that pseudo processes are not counted as causal processes (as Salmon admits in Salmon 1997: sec. 2). Salmon regards the absence of counterfactual requirements to be the major improvement of this theory over the mark theory. On the other hand, perhaps the most significant difference from the CQ theory is the requirement that the quantity be *transmitted*. This involves two significant features. First, it brings in a causal direction, the direction of transmission; and second, it requires causal processes to be spatiotemporally continuous. I shall discuss these aspects in the following sections. In the final section I shall also comment on Salmon's preference for 'invariant quantities' over 'conserved quantities.'

V.6.2 Transmission and the Direction of Causation

Although Salmon does not explicitly discuss the issue, the IQ theory appears to express a directionality not found in the minimal CQ theory.

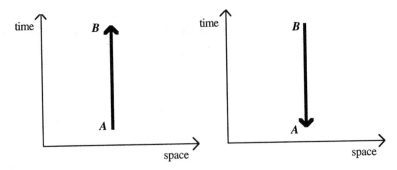

Figure 5.4. Backwards transmission.

This appears in IQ3, which is directional in that it involves the phrase *'from A to B.'* This direction of transmission presumably is to be identified as the direction of causation, but should not be confused with the direction of time. Salmon wants his theory to be in principle independent of time, so it is appropriate that the IQ theory allows the possibility of backwards-in-time causation, in the following way. Suppose *B* occurs later than *A*, and that some invariant quantity is transmitted from *B* to *A*. Then, for this particular process, the causal arrow points in the opposite direction to the temporal arrow, that is, we have a case of backwards-in-time causation.

However, we need to look more closely at the notion of the *direction* of transmission. How is it to be understood? IQ3 tells us that transmission of a quantity from *A* to *B* amounts to possession, by the object in question, of that quantity at each spacetime point over the relevant interval (the 'at-at' theory of transmission). But notice that the left-hand side of this definition has directionality – 'from *A* to *B*'; but the right hand-side does not – 'possession at each spacetime point.' This means that transmission, once defined, turns out to lack directionality. Thus the account appears to offer an account of causal direction, but in fact fails to deliver the goods.

The point may be brought out as follows. Consider two causal processes with opposite directionality: for one, the direction of causation is normal, and for the other, the direction of causation is backwards-in-time (see Figure 5.4). Now, the IQ theory does not distinguish between these processes, for both possess the relevant quantities at each spacetime point between *A* and *B*. In other words, a process will possess the relevant quantity at each point regardless of whether it transmits that quantity from *A* to *B* or from *B* to *A*. In cases such as the backwards-in-time causality posited in quantum mechanics (see Chapter 8) the reasons for judging the direction of causation to be backwards are much more complex than just the continuous manifestation of some quantity; in fact, it is a highly theoretical matter (see Dowe 1992b: sec. 4). Thus IQ3 does not provide an account of the direction of causation. Indeed, the notion of 'transmission' turns out to involve no more than what is contained in the notion of 'possession.' I conclude that there is no advantage in appealing to 'transmission' (as in the IQ theory) rather than to 'possession' (as in the CQ theory). Thus Salmon has a problem similar to that of Aronson,[18] except that Salmon has spelt out the intended meaning of 'transfer,' making its failure more perspicuous.

Salmon used to have the resources for an account of causal direction, when conjunctive forks played a central role in his account of causal processes and interactions. That resource is the contingent fact that forks are open to the future but not to the past. For a number of reasons, Salmon no longer appeals explicitly to that fact in his explication of causality (Salmon 1990; 1994: 301). But that is not to say that it could not feature as an *additional* factor to explain causation's direction. (For a discussion of how that might be done, see Chapter 8.)

Since transmission involves no more than continuous possession, I suggest that IQ2 be replaced by the following:

IQ2′. A *causal process* is a world line of an object that possesses a constant amount of a nonzero invariant quantity at each moment of its history.

This renders IQ3 superfluous. Then the IQ theory is minimalist with respect to asymmetry in much the same way that the CQ theory is.

V.6.3 Miscellaneous Differences: Which Quantities and When?

In this section various remaining points of difference between the CQ theory and the IQ theory are discussed. First, IQ2′ requires that a causal process possess a *fixed amount* of the relevant quantity, in the absence of interactions. Of course, if we are talking about conserved quantities, then this obviously must be true. However, there are pragmatic difficulties that we have already encountered: actual causal

18. See Chapter 3, section 3.4.

processes do not operate in the absence of interactions. A body moving under the action of a field, for example, is really a string of interactions. In reality most causal processes attenuate, as there is loss of energy to the environment. It seems that if we wish to accommodate the standard types of causal processes – for example, Salmon's example of a moving baseball – then we will have difficulties with this requirement. In theory, causal processes do possess a fixed amount of energy, say, in the absence of interactions, but in practice such a requirement renders the notion of a causal process, as opposed to an interaction, useless. For this reason the CQ theory does not require that a causal process possess a constant amount of the relevant quantity over the entire history of the process.

Second, Salmon requires that the amount possessed be *nonzero*. In his discussion of the CQ theory, Salmon notes a "possibly serious ambiguity" (1994: 306). An object might possess no energy, let's say, because it is currently at rest, although it is the type of object that can possess energy, and perhaps does at other times. This is different from an object, such as a shadow, that cannot possess energy. "To be on the safe side," Salmon requires that the object possess a nonzero amount of the relevant quantity. I think this is the wrong option. The world line of an object at rest possesses zero momentum, but is a causal process. Such an object may enter into a collision with a moving object, in which case its momentum (zero) will enter into the equation. Shadows, on the other hand, do not enter into the equation. Shadows and balls at rest differ in that the former are not the type of objects to which conserved quantities may be ascribed, whereas the latter are. This seems to mirror the fact that the former are pseudo processes, while the latter are causal processes. So the CQ theory is to be read as allowing causal objects to possess a zero amount of the quantity in question.

Third, Salmon requires that a causal process possess the relevant quantity *at each moment* of its history. This seems fair enough – it only demands that *if* the object exists at a time, then it must possess the quantity at that time. If an object ever possesses a conserved quantity, then it will always possess that (perhaps varying amount, perhaps zero) quantity. The CQ theory omits this condition only because it is superfluous, not because it is false. I take it that this difference is relatively innocuous.

However, an apparently similar condition is not so innocuous. If the requirement is that the object possess the quantity at every moment *of the relevant interval*, then the notion of spatiotemporal continuity has been introduced. The former consideration did not require that the object exist at every instant over the interval, but this requirement does. It therefore rules out the possibility of a causal process being gappy, or discontinuous. As will be clear in Chapter 8, I happen to prefer the backwards causation solution of Bell's Theorem, but it would be a mistake to rule out a priori the solutions involving other types of nonlocality. For this reason it is good that the CQ theory involves no commitment to spatiotemporal continuity (see Forge 1982).

Finally, we turn to the question, which quantities? Salmon argues that invariant, not conserved quantities are the appropriate ones, because causality is invariant: "if two events are causally connectable in one frame of reference, they are causally connectable in every frame" (Salmon 1994: 305). However, Salmon also notes an immediate difficulty: all spacetime intervals are invariant, and pseudo processes involve spacetime intervals. One possibility, then, would be to require that causal processes possess invariant conserved quantities. Salmon himself attempts to avoid the difficulty by appeal to the notion of transmission. I have already discussed why I reject that appeal, but the solution seems tenuous on other grounds as well. Salmon's idea is that a shadow, for example, although it possesses the invariant quantity, does not possess it at each spacetime point, and therefore cannot be said to transmit it. This appears, at the very least, unclear. It seems to me that one might just as easily have insisted that the shadow possess its spacetime interval at every spacetime point.

I have no strong objection to the requirement that causal processes possess invariant conserved quantities. This entails a minor modification to the CQ theory. However, I don't think the requirement is necessary. It is true that an object's linear momentum varies depending on the frame of reference, but the CQ theory is not concerned with how much momentum the object possesses, just whether it is the type of object that possesses such a quantity. *That* the object possesses momentum is a fact that does not vary with the frame of reference. The concept of an exchange of momentum also is invariant, since change in momentum is invariant.

V.6.4 Salmon's Second Revision

In his 1997 *Philosophy of Science* paper Salmon has responded to these criticisms, and offered a further revision to the theory. Salmon accepts the criticisms just presented about the direction of causation, and about invariant quantities. However, he is not willing to replace the notion of 'transmission' with 'possession of conserved quantity by an object that exhibits genidentity.' He therefore presents the following revised theory of causality:

A causal process is the world line of an object that transmits a non-zero amount of a conserved quantity at each moment of its history (each spacetime point of its trajectory). (1997: 468)

The concept of transmission is to be understood by the following definition.

A process transmits a conserved quantity between A and B (A ≠ B) if and only if it possesses [a fixed amount of] this quantity at A and at B and at every stage of the process between A and B without any interaction in the open interval (A, B) that involves an exchange of that particular conserved quantity. (1997: 463)

Thus, instead of adopting the assumption about identity, Salmon appeals to a special kind of regularity that involves the possession of a fixed amount of a conserved quantity at every spacetime point of the process; in other words, another version of the 'at-at' theory.

V.6.5 Noninteracting Intersections

I want to argue that Salmon's account does not avoid the presumption of identity. The problem I want to focus on involves non-interacting intersections. For example, suppose we have two particles that are able to pass through each other without having any effect on each other. Then, on Salmon's theory, we are not able to tell whether we have a case where two causal processes pass through each other or a causal interaction where two particles collide and exchange quantities on colliding. But the assumption of identity over time determines which of these two alternatives is in fact the case.

Suppose we have two objects exactly similar, except that one has conserved quantity $q = n$, while the other has $q = m$ ($m \neq n$). Suppose, as before, they pass (or appear to pass) straight through each other, so that their world lines intersect. After the intersection there is no change to the nonspecified properties of either. Is this a causal interaction?

Both Salmon and I now agree that a causal interaction involves an exchange of a conserved quantity. However, Salmon's account of a causal process fails to decide unambiguously whether our case is a causal interaction. One alternative has it that object a ($q = m$) and object b ($q = n$) pass by each other unchanged: no causal interaction. But another alternative says that there is a causal interaction where a ($q = m$) changes direction and amount of q (to q= =) while b also changes direction and amount of q (to $q = m$). This involves an exchange of a conserved quantity, and so qualifies as a causal interaction. Disambiguation is achieved by the notion of identity through time: if it is the case that a is the same object before and after the intersection, then there is no causal interaction.

This is not such an uncommon scenario. A stray neutrino passes through my body. Am I still myself, unaffected by the event, or am I now the thing that used to be the neutrino, having been radically transformed by the experience? Other noninteracting intersections include radio waves entering a building and light coming through the glass window.

Salmon's theory requires, in a causal process, that a conserved quantity be possessed continuously, in the absence of causal interaction. But in noninteracting intersections it would be indeterminate as to whether we have a causal process.

So, Salmon's theory does require a notion of identity through time. Continuous possession does not avoid the problem. At least we may say, on a pairwise consideration of Salmon's and my own theory, that I have answered Salmon's claim to have uncovered a reason to prefer his to mine.

V.7 SUMMARY

In this chapter a minimal physical account of causation has been presented, which explicates the core notions of 'causal process' and 'causal interaction.' I have argued that the present theory does better than its rivals in characterising these core notions. The account is minimal in the sense that it is not committed to dubious notions of transfer and transmission. As a minimal account, it does not fully account for every aspect of 'causation' as we normally use the word. In Chapter 7 we will consider the connection between causal processes/interactions and the relation of cause that we think obtains between facts or events. In Chapter 8 we will consider the asymmetry of that relation. I do not wish to be committed a priori to any view about whether these additional connotations of the word have legitimate grounding 'in the objects.' Before we turn to these questions, however, there is another objection that needs attention.

6

Prevention and Omission

In this chapter we address an issue that is a difficulty for the Conserved Quantity theory; and not only for the Conserved Quantity theory, but also for many approaches to causation. We offer a solution that is available not only for the Conserved Quantity theory, but for most theories of causation.

We might be tempted to think that preventions, such as 'the father's grabbing the child prevented the accident,' and cases of causation by omission, such as 'the father's inattention was the cause of the child's accident,' are examples of causation. Such cases are 'causation' by *prevention* or *omission*, and they almost always involve negative events or facts as one or both of the relata. I will call this relation 'causation*.'

For example, in 'the father's inattention was the cause of the accident,' the effect is a real occurrence, but the cause is an omission, a failure to do something. On the other hand, in 'his grabbing the child prevented the accident,' the cause is a real occurrence but the effect is the nonoccurrence of something. The former is omission, the later prevention, but in both cases we have causation*, and both cases involve negative events or facts. In both cases, we can recognise that it is not literal causation, yet we still want to use the term 'cause.'

In general, suppose we have a case of singular causation* where A causes* B. If A is a negative event or fact, then we have a case of causation* by omission; if B is a negative event or fact, then we have a case of causation* by prevention. If both A and B are negative events or facts, then we have a case of prevention by omission. The problem is how negative events can enter into real causal relations. How can anything cause an event to not happen? Or how can something that doesn't exist actually cause anything?

The problem is particularly acute for the Conserved Quantity theory. There is no set of causal processes and interactions linking the father's inattention to the accident, or the father's grabbing the child to the nonoccurrence of the accident. A particularly perspicuous example is due to Brian Ellis: pulling down the blind causes the room to be dark, but the light is prevented from entering the room.

The problem of prevention and omission is in fact an acknowledged problem for most theories of causation. I claim it can be solved in a way that is consistent with all theories of causation. My claim is that causation* should be understood not as real causation but as a hybrid fact usually involving certain actual real causation together with certain counterfactual truths about real causation, but that nevertheless we are justified in treating such cases as causation for practical purposes. This chapter is an articulation and defence of that claim. In other words, I am offering a counterfactual theory of prevention and omission.

To my knowledge the only philosopher to have defended this intuition in print is David Fair (1979: 246–247). However, Fair's account is given in terms of 'plausible possible worlds,' which makes preventions and omissions relative to human beliefs. I take preventions and omissions to be an objective matter.

In the first section, The Intuition of Difference, I identify a certain intuition that supports my case. In the second section, A Universal Problem, I show how omissions and preventions are a problem not only for the Conserved Quantity theory, but also for a broad range of theories of causation. In section 6.3, I outline the role that negative events or facts play in omissions and preventions, and argue that the problem is much more widespread than is usually recognised. In sections 6.4 and 6.5, I offer an analysis of preventions and omissions, respectively. Section 6.6 deals with more complex cases. Finally, in section 6.7, I show how this analysis is a 'cross-platform' solution, that is, it is compatible with a range of major theories of causation.

VI.1 THE INTUITION OF DIFFERENCE

My main argument for the counterfactual theory is that it solves the problems that prevention and omission cause not only for the Conserved Quantity theory, but also for a wide range of theories of causation. However, it will prove helpful to have an independent motivation for drawing a distinction between prevention or omission and genuine causation. What I will call the 'intuition of difference' provides such a motivation.

In the right sort of case, we have an intuition that there is a difference between prevention and causation by omission on the one hand, and genuine causation on the other. Consider the following dialogue:

You say that the father's inattention was the cause of the child's accident. Surely you don't mean that he literally made the child run into the path of the car, or that he made the car hit the child. Rather, you mean that his failure to guard the child was the cause in the sense that if he had guarded the child, the accident would not have happened. You don't mean that he literally caused the accident; you mean that it was possible for him to have prevented it.

Yes, that's what I mean. And you, when you talk of a scenario in which the father prevented the accident, you don't mean that he bore any literal causal connection to a real thing called an accident. You mean that had he not acted in the way that he did, some circumstances would have brought about the accident.

Yes, that's exactly what I mean.

I claim that we can recognise, on reflection, that certain cases of prevention or omission, such as this one, are not really cases of genuine causation. Call this the 'intuition of difference.' We also feel, however, that the 'mistake' of treating them as if they were causation doesn't matter for practical purposes.

There is further evidence for the existence of this intuition in the fact that the problem of omission and prevention arises in a large variety of philosophical contexts. In the literature on the causal theory of perception there has been debate about whether we see black holes (see Goldman 1977: 281–283; Tye 1982: 234). In the literature on euthanasia there is the question of the distinction between killing and letting die, and of whether allowing a death is to cause it (Bennett 1995; Glover 1977: chap. 7). In philosophy of law there is the issue of whether neglect is an example of causation (Hart and Honore 1985: 140; Mullany 1992). The mere fact that the distinction is noticed and debated is proof enough that there is an intuition of difference.

I am not saying that the intuition of difference is a particularly strong intuition, nor that there are no conflicting intuitions, nor that such intuitions are inviolatable givens in philosophical theorising. At this stage I claim only that under the right conditions we can recognise that we do have such an intuition.

The central argument of this chapter does not rely on the intuition of difference. The central argument is that the counterfactual account of prevention and omission is able to solve the problems that these cause for theories of causation. We now turn to a survey of these problems.

Prevention and causation by omission create (mostly) acknowledged problems for a quite disparate range of theories of causation. In this section I will briefly survey some problems caused for the Conserved Quantity and the transference theories, and also for the theories of Lewis, Armstrong, Suppes and others.

The Conserved Quantity theory holds that causation is primarily about the set of causal processes and interactions that constitute the causal structure of the world. How this set of processes and interactions connects the things we think of as 'causes and effects' remains to be seen (Chapter 7). However, it is reasonable to suppose that it is a necessary condition for two events to be linked as cause and effect that they be connected by a set of causal processes and interactions.

This is sufficient to cause trouble with preventions and omissions. Clearly there is no set of causal processes in this sense that can be traced out between the father's grabbing the child and the accident not happening, or between the father's inattention and the accident. These events are not linked by a set of causal processes and interactions. Either omission and prevention are not causation, or the Conserved Quantity theory of causation is false (see Hausman 1998: 15–16).

Preventions and omissions are just as much a problem for the transference theory (Aronson 1971b; Fair 1979), according to which causation is the flow of energy or momentum from one object to another. As Ehring (1986) points out, there are cases we call causation where there is no such transference, such as 'flicking the switch causes the light to go out.' Ehring calls this causation by 'eliminating transference,' and Aronson classes such events as 'occasions' rather than causes, but I think that this is a clear case of preventing: flicking the switch prevents the light from remaining on. In any case, the flicking of the switch and the light being off are not linked by a transference of energy. Our case of the father's neglect and the accident is even clearer: there simply is no transference of energy or momentum from one to the other. Either omission and prevention is not causation, or the transference theory of causation is false. This is equally a difficulty for the 'force' theories of Bigelow, Ellis and Pargetter (1988), Bigelow and Pargetter (1990a) and Heathcote (1989).

However, the problem is by no means restricted to this kind of account. First, allowing negative events as satisfactory candidates for the relata of causation causes problems for David Lewis's counterfactual theory of causation, as Lewis himself has shown (1986: 189–193). According to that theory there is causal dependence between occurrent events A and B just if had A not occurred, B would not have occurred (1986: chap. 21). On Lewis's account an event must not be too disjunctive (e.g., 'my going to the movies or having a car accident' is too disjunctive to count as an event), one reason for this being that an effect counterfactually depends on any disjunctive event that contains its cause as one of the disjuncts (Lewis 1986: 267).

If we accept omissions as genuine events essentially specified as omissions (explicated as 'the father was tying his shoelace or intently gazing into the sky or . . .'), then they fail the requirement that events not be highly disjunctive. If instead we count the event that caused the accident – the father's lack of attention – as just the actual things he did (say, tying his shoelace), then there is no longer a problem with disjunctive events, but there is a problem with counterfactual dependence, since it is not true that had the father not done those actual positive things he did, the accident would not have occurred; since there are innumerable ways that the father could have failed to prevent the accident.

The following objection is mistaken. The same problem arises for positive events, it might be argued. Events can have different degrees of fragility, so the child's running onto the road (X) could be taken as the child's running with her hands clenched (X′), a more fragile version of X; and if we do, then there's trouble for the counterfactual account of causation since the following is false: 'had X′ not occurred, the accident would not have happened'; therefore, to save counterfactual dependence Lewis has to take positives as disjunctive events as well: the child's running this way or the child's running this slightly different way, and so on. But the mistake here is to miss a crucial difference between the positive and negative cases: the positive disjunctive event is appropriately unified by similarity and so is not highly disjunctive, whereas the negative is not, and so is highly disjunctive; and it is highly disjunctive events that must be ruled out by the earlier reasoning.[1]

Taking facts instead of events as the appropriate relata also presents problems. According to David Armstrong, the world is a world of states

---

1. It may be that counterfactual accounts that deal with these problems in different ways (e.g., Ramachandran 1997) will be able to handle negatives. My point here is simply that Lewis's theory cannot account for negatives.

of affairs in which singular causation, conceptually primitive, is a second-order relation between first-order actual states of affairs (such as a certain particular having a certain property) (Armstrong 1997: chap. 14).[2] However, there are no negative states of affairs (1997: 134–135); the truthmakers for true negative statements include higher-order 'totality' states of affairs, such as 'such and such are the only states of affairs.' So omissions and prevented states of affairs cannot be the relata of causation because they are not, and do not supervene on, actual first-order states of affairs.

There are accounts of causation according to which cases of omission and prevention come out as clear cases of causation. For example, take Suppes's probabilistic theory of causation (1970). According to Suppes, C is a prima facie cause of E just if C occurs before E and the conditional probability of E occurring when C occurs is greater than the unconditional probability of E occurring (Suppes 1984: 48). For Suppes an event is set-theoretical – anything to which a time and a probability can be assigned – so a negative event is as good a candidate as a positive event. The father's grabbing the child lowers the probability of the accident, and his failure to do so raises it. But such theories fail to give an adequate account of omissions and preventions in the sense that they fail to account for the 'difference intuition.' As is evident in the dialogue presented in the previous section, in certain clear-cut cases we do recognise that omissions and preventions are not *literally* cases of causation.

The same can be said for manipulability theories of causation such as those due to Collingwood, Gasking and von Wright (see also Mellor 1995: chap. 7 and Price 1996b: chap. 7), according to which causes are essentially means for bringing about their effects as ends. Each of these authors treats prevention as a straightforward case of causation (Collingwood 1974: 119; Gasking 1996: 111, 114; von Wright 1971: 69–72). Negative events can be ends and (apparently) means. Guarding the child would be the means for bringing about the desired end, the nonoccurrence of the accident. Collingwood adds a note that preventing something is producing something incompatible with it (1974: 119). Von Wright comes closest to considering the case of omission, except that he seems to be thinking purely in terms of causal laws:

---

2. See also Mellor (1995).

p is a cause relative to q, if and only if by doing p we could bring about q or by suppressing p we could remove q or prevent it from happening. (1971: 70)

Although nothing in this book hangs on it except the exact scope of the claim that omissions and preventions are a universal problem, I want to urge that these sorts of theories also fail to give an adequate account of omissions and preventions. The problem is simply that they fail to account for the fact that we do recognise that omissions and preventions are not *literally* cases of causation, although for some reason it's OK to treat them as causation. This is evident in the dialogue cited earlier: when we say that the father's inattention was the cause of the child's accident we don't mean that he literally made the child run into the path of the car, or that he made the car hit the child; and when we say that the father prevented the accident, we don't mean that he bore any literal causal connection to a real thing called an accident. Theories such as Suppes's simply allow omissions and preventions as genuine causation. But far from thereby solving the problem, they actually fail to account for the fact that we recognise that omissions and preventions are not causation, strictly speaking.

This survey of some of the problems facing contemporary theories of causation suffices to show that the problem of negatives in causation is a widespread and significant problem for a range of contemporary approaches to causation. This is not to claim that there is no theory that has a solution to any of these problems. I simply claim that there is a range of significant and quite different theories of causation that do have a problem, and that the counterfactual theory of prevention and omission solves it, and does so in a manner that makes the solution compatible with that broad range of theories.

In the next section I show how widespread the problem is, in the sense that it affects more cases of 'causation' than we may at first recognise. In order to show this it will first be necessary to show in more detail how negatives are involved in causation by omission and prevention.

VI.3 CAUSATION AND NEGATIVE EVENTS

Most cases of omissions and preventions involve negative events or facts. In the simplest case, an omission is where a negative event causes something; prevention is where something causes a negative event.

Now there is a certain linguistic conventionality about negative events, since for any events A and B, if A is not-B, then B is not-A. To avoid an accident is to not have an accident, to have an accident is to not avoid an accident. Any event given in the negative can be rewritten in the positive, provided one can think of the right word.

But for there to be a real difference between omissions and preventions on the one hand, and causation on the other, there must be a real distinction between positive and negative events. Some names for events name positive events, some name negative ones.

For factualist accounts such as Armstrong's (1997) and Mellor's (1995), real facts (states of affairs for Armstrong, facta for Mellor) involve particulars having genuine properties, which are universals.

According to the Conserved Quantity theory, cause and effect are connected by a set of causal processes and interactions, where a causal process is the world line of an object that possesses a conserved quantity, and a causal interaction involves an exchange of a conserved quantity. If such quantities (energy, momentum, charge etc.) are the genuine properties, then the genuine positive events or facts involve the possession of those quantities. In Ehring's case where flicking the switch causes the light to go out, 'the light being out' is a negative event, the filament not radiating energy (1986: 251). In Ellis's case where the pulling down the blind causes the room to be dark, 'the room being dark' is a negative fact, its not being illuminated. In our case of the accident, having an accident is a genuine causal interaction, involving the exchange of momentum, whereas not having an accident is not a genuine causal interaction, since it involves no exchange of a conserved quantity.

Of course, the thesis that there is a real distinction between negative and positive events is not a thesis tied to the Conserved Quantity theory. As already mentioned, it is a distinction entailed by the distinction between omissions and preventions on the one hand, and genuine causation on the other.

Nevertheless, it may not always be clear whether an event or fact in question is positive or negative. Events we think of as negative may turn out really to be positive. In those cases, apparent omissions and preventions turn out to be cases of ordinary genuine causation.

Alternatively, apparently positive events may turn out to be negative events, and consequently, cases of apparently genuine causation may turn out to be omissions or preventions. The latter is especially convincing, as Lewis and Armstrong have pointed out to me. Lewis gives the case of drowsiness caused by opium. This appears to be genuine causation between two positive events, but it may be that opium actually prevents some normal process from performing its normal function.

But these considerations are of merely epistemic concern. We may not know whether a given case is a prevention or causation, but the conceptual distinction between genuine causation and omissions/preventions is clear enough; just as we may be unable to tell whether a given star is a binary star or a pulsar, when the theoretical distinction is perfectly clear. On the other hand, this possibly widespread uncertainty shows why it is handy to treat causation and causation* as if they were the same thing. This is exactly what we do for practical purposes, as we have seen.[3]

These considerations illustrate the claim that preventions and omissions may be very much more commonplace than is at first recognised. If so, there is serious need to deal with the issue as part of any serious theory of causation. Take some common examples of causation: "Drink-driving causes accidents." But perhaps alcohol prevents normal functioning (a prevention), which in turn causes the accident (causation by omission). "Smoking causes heart disease." But perhaps the effect of tobacco is to prevent normal processes from impacting certain cells in a certain way, so that, in the absence of those processes, diseased cells prosper (causation by omission).

Preventions and omissions are widespread in another way. When a single positive cause produces one positive effect, it often (if not always) also produces infinitely many negative effects (preventions). So, the force F applied by the cue stick is causally responsible for the magnitude of the subsequent velocity s of the cue ball. But at the same time, that same force F is also causally responsible (in the sense of a prevention) for the fact that the cue ball does not have speed s' or s'' or s''' (where each is $\neq s$), and an infinite number of other preventions (compare Mellor 1995: 51).

The first step in analysing omissions and preventions is to describe the case in terms of genuine positive events or the negation of positive events, as the case may be; that is, to write in the form of 'A' for genuine

---

3. Furthermore, as Armstrong has pointed out, the epistemic blur may explain why some folk and metaphysicians have failed to see the distinction.

positive events, and 'not-A' for genuine negative events. This may reveal that the apparent prevention is a case of genuine causation, in the case that the effect is really the negation of a positive event. It may also reveal that cases expressed as straight causation are really preventions or omissions. Thus 'pulling down the blind caused the room to be dark' (Ellis) is really a prevention, 'the room being illuminated' being the positive event.

So for the purposes of analysis we will take the canonical forms for genuine omissions and preventions to be:

Prevention: A caused* not-B
Omission: not-A caused* B

where A and B are genuinely positive events. This, of course, does not tell us in what sense negative events can be causes or effects. We turn to that question in the next two sections.

VI.4 PREVENTION

We have a case of prevention just if A causes* not-B, where B is a positive event or fact. For example, the father's grabbing the child prevented the accident, that is, caused* the accident not to happen. My claim is that this should be understood not as genuine causation but as a counterfactual truth about genuine causation. In this section I spell out what is meant by that. The account of prevention I propose is as follows:

A prevented B ≡ A caused* not-B if

(P1) A occurred and B did not, and there occurred an x such that
(P2) there is a causal relation between A and the process due to x, such that (as in Figure 6.1) either
  (i) A is a causal interaction with the causal process x, or
  (ii) A causes y, a causal interaction with the causal process x,
and
(P3) if A had not occurred, x would have caused B,

where A and B name positive events or facts, and x and y are variables ranging over events and/or facts. In section 6.6 we will see why the account only provides a sufficient condition. Although as far as I can see for the present purposes nothing in particular hangs on it, the coun-

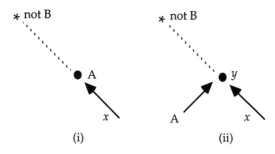

Figure 6.1. (i) A is an interaction involving the process x. (ii) A causes y, an interaction involving the process x.

terfactuals here and elsewhere in this chapter will be understood in the fashion of David Lewis (1986), in the sense that they will be taken as closest-world conditionals, with no backtracking, and so on. All we really need in that regard is acceptance of the idea that there are counterfactual truths about causation. If there are, then whatever is the appropriate semantics for them will be the right semantics here.

The term 'causation' as it appears in (P2) and (P3) refers to genuine causation, not causation*. This means that genuine causation is primitive with respect to this theory of preventions, the implication being that one may insert one's own theory of causation, whatever that may be. (I will return to this point in section 6.7.)

It is important to underline the difference between the present account and Lewis's counterfactual account of causation. Lewis uses counterfactuals to 'analyse out' causation, whereas this account uses counterfactual claims about genuine causation to analyse preventions. Causation itself is retained in the counterfactual.

The account captures the sense, when A prevents B, that although A doesn't literally cause not-B, there is nevertheless some significant relation between A and not-B. This 'relation' is a hybrid one, involving some actual genuine causation and a counterfactual truth about genuine causation. The actual genuine causation is usually that A caused something – according to (P2) – and the counterfactual truth about genuine causation is that if A hadn't happened, something would have caused B (P3).

The most common case is (ii), where the preventing event A causes an interference in some process that would otherwise lead to B.

133

However, (i) is also needed not only because sometimes the event we say prevents B is simply the interruption of process x that would have caused B (the fact that the car slowed down prevented the accident), but also because in normal cases where the preventing event A causes an interference in x (e.g., the father's grabbing the child prevented the accident), we would also be happy to say that the event y – the interruption of process x – itself also prevents B (the fact that the child was stopped from running towards the road prevented the accident). The preventer A can *cause* the interruption, or it can *be* the interruption.

How A prevents B is not part of the meaning of the claim 'A prevented B,' but in fact it was by one of ways (i) or (ii), as depicted in Figure 6.1. Both cases involve A in actual causation. This is important, because although how A prevents B is not part of the meaning of 'A prevented B,' it clearly *is* part of the meaning that A is somehow responsible for the fact that B didn't occur. (In section 6.6 we will see that there is one exception to this – the case of prevention by omission.)

We need to say more to account for alternative preventers. There are two types, preemptive prevention (cf. preemption) and overprevention (cf. overdetermination); in both cases (P3) fails. First, suppose the father grabs the child, preventing the accident. But suppose had the father not done so, the mother would have grabbed the child. Then it is not true that, as per (P3), if A had not occurred, x would have caused B, but it is true that had neither A nor C (the mother grabbed the child) occurred, x would have caused B. C will not itself qualify as a preventer of B because it does not occur. Second, suppose the father grabs the child, preventing the accident. But suppose that at the same time the mother shouts (C), causing the child to stop running, also preventing the accident. Then for neither event does (P3) hold. Even had the father not grabbed the child, the child's running towards the road would not have caused the accident; and even had the mother not shouted, the child's running towards the road would not have caused the accident. But it is true that had the father not grabbed the child *and* had the mother not shouted, the child's running towards the road would have caused the accident. We need to disjoin (P3) with

(P3') there exists a C such that had neither A nor C occurred, x would have caused B or . . .

So by (P3′) A prevents B. It is also true that C prevents B, so we have overprevention.[4] This approach can be iterated for any number of alternative preventers, by adding further conditions as alternatives to (P3).

We now turn to an extended example. Suppose:

(a) The father's yelling to the child prevented the accident.

The events A (the father's yelling) and B (the accident occurred) are both positive events. For the father's yelling to the child to be something that prevented the accident, it must cause something; in this case, suppose it caused y (that the child stopped running in the direction of the road). It is not part of the meaning of 'A prevents B' that A caused this particular event, but in the actual scenario, it did. Then it is also required that had A not occurred in those circumstances, then the fact that the child was running onto the road would have caused an accident. This case is an example covered by (P2 i), depicted in Figure 6.1(i).

Or suppose instead:

(b) The father's grabbing the child prevented the accident.

The events A (the father's grabbing the child) and B (the accident occurred) are both positive events. For the father's grabbing the child to be something that prevented the accident, it must cause something; here it ends the process x, the child's running in the direction of the road. Again, had A not occurred in these circumstances, then the fact that the child was running onto the road would have caused the accident. This case is an example covered by (P2 ii), and depicted in Figure 6.1(ii).

What seems to happen, at least in these sorts of preventions, is that the prevented event, considered in its possibility, is an interaction between two causal processes. In our example, the accident is an interaction between two causal processes, the child and the car. The preventing event, the father's grabbing the child, interrupted one of these processes, thereby preventing the accident. The actual causation involved is the causal interaction between the father and the process that would have led to the accident, namely the running child.

4. This solution bears some resemblance to one offered by Ganeri, Nordoff and Ramachandran (1996).

We have a case of causation* by omission whenever not-A causes* B, where A and B are positive events or facts, and not-A is the 'act of omission.' The father's failure to guard the child caused* the accident. Omissions are, like preventions, not cases of genuine causation but, again, a counterfactual claim about genuine causation. The account is as follows:

Omission: not-A caused* B if

(O1) B occurred and A did not, and there occurred an x such that
(O2) x caused B, and
(O3) if A had occurred then B would not have occurred, and there would have been a causal relation between A and the process due to x, such that either
   (i) A is a causal interaction involving the causal process x, or
  (ii) A causes y, a causal interaction involving the causal process x,

where A and B name positive events, and x and y are variables ranging over facts or events. This is another way of saying that had A occurred, it would have prevented B.

It may be that in the circumstances there lurks a back-up cause, besides x, such that even if A had occurred, B would still have occurred. This doesn't change the analysis, because not-A still causes* B by omission. The father's failure to warn the driver still causes* the accident even if, had that driver slowed down, some other car would have hit the child. There is an analogy here to genuine causation: the speeding car caused the child's death even if, had its driver pulled over, some other car would have killed the child.

Although causation* by omission is not genuine causation, it does involve some actual genuine causation, namely that something caused B. It also involves a counterfactual about genuine causation, that had the omission occurred, the effect B would not have, and further, that it would not have because A would have interfered with the process that actually brought about B. It is on the strength of this that we are justified for practical purposes in talking about causation* by omission as if it were genuine causation.

As an example of case (i), suppose:

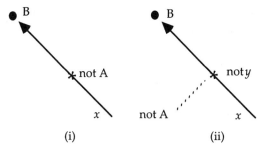

Figure 6.2. x causes B, but if A had occurred, (i) A would have been an interaction involving process x, (ii) A would have caused y, an interaction involving process x.

(a) The father's failure to grab the child caused* the accident.

Then x, the child's running onto the road, actually caused the accident. Also, as described in (O3 i), the father's grabbing the child, if it had happened, would have been the interruption of the process that actually led to the accident, namely, the child's running towards the road. An example of case (ii) is:

(b) The father's failure to warn the child caused* the accident.

Again, x, the child's running onto the road, actually caused the accident. But here, as described in (O3 ii), the father's warning the child, if it had happened, would have caused the interruption of the process that actually led to the accident, namely, the child's running towards the road.

For an omission to cause* its effect, the possible causation by virtue of which it does so must be possible, in some appropriate sense. It can't be 'logically possible,' since by the Humean doctrine of causal independence anything can cause anything. The obvious candidate is physical possibility. Then we need a further restriction on what kinds of events can be omissions. We think it is physically possible for me to call out to the child, and so my failure to do so caused* the accident. But it is not physically possible for a stone to cry out, so it should not be that the stone's failure to call out to the child caused* the accident. Hence we restrict the events whose negation can be omissions to physically possible events.

Omissions bear an interesting relation to preventions. Notice that in our definition of causation* by omission we could have added two superfluous conditions: first, that had A occurred, A would have occurred, and second, that had A occurred, then had A not occurred x would have caused B. If we wrote these conditions into clause (O3), then, putting together the definitions of prevention and omission, we would have the result that not-A caused B by omission if and only if something (x) caused B and had A occurred, A would have prevented B (by preventing the x that actually caused B).

In fact, in the right circumstances, the following mirror relation holds between prevention and omission: Where a preventer A caused* not-B, had A not occurred, not-A would have caused B by omission; and where an omission not-A caused* B, had A occurred, A would have prevented B. In other words, had a preventer A (which caused* not-B) not occurred, its nonoccurrence would have been an omission (which caused* B); and had an omission not-A (which caused* B) occurred, its occurrence would have been a preventer (which caused* not-B). Thus there is a mirror relation between preventions and omissions, as in Figure 6.3. In both cases, the scenario we imagine in the all-important counterfactual is just the mirror case.

Thus the underlying structure of actual and counterfactual causation that makes statements about prevention and omission true also makes true this mirror relation between prevention and omission. This relation between 'A caused* not-B' and 'not-A caused* B' has a correlative in genuine causation. In most circumstances, for deterministic cases, 'A caused B' seems to entail that if A had not occurred, not-A would have caused not-B. If John's drinking arsenic caused his

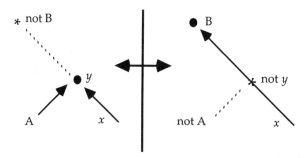

Figure 6.3. A mirror: prevention and omission.

138

death, then had he not drunk arsenic that would have prevented his death.

There is one blemish in the mirror as I've presented it. In the definition of preventions we left the case where there are alternative possible preventers as a special case, whereas in the case of omissions we accounted for them, leaving out a clause to the effect that had A occurred, x would have caused B.

We have so far only considered cases where not-A caused* B and A caused* not-B. This suggests a further case where not-A caused* not-B, where the omission causes* a nonoccurrence. In fact, in such a case we have *prevention by omission*, but there are at least two distinct types of cases. In the first type, the analysis is simple:

Prevention by omission, type 1: not-A caused* not-B if

(1) neither A nor B occurred, and
(2) if A had occurred, A would have caused B,

where A and B name positive events or facts.

This is the case where an omission prevents something simply by not causing it. So, for example, 'John's not drinking arsenic prevented his death,' is true in the circumstances if the counterfactual 'if John had drunk arsenic, his drinking arsenic would have caused his death' is true – a counterfactual about genuine causation, as in Figure 6.4 (i). This is the case, alluded to earlier, where there is no actual causation present.

The second type of prevention by omission is not so simple:

Prevention by omission, type 2: not-A caused* not-B if

(1) neither A nor B occurred, and there occurred an x, a y, and a z such that
(2) process z caused y, an interaction involving the process due to x, and
(3) had A occurred, then
   (a) y would not have occurred, B would have occurred, and there would be a w such that

139

(b) A would have caused *w*, an interaction involving the process due to *y*, and

(c) the process due to *x* would have caused B,

where A and B name positive events, and *w*, *x*, *y* and *z* are variables ranging over facts or events.

This type of prevention by omission occurs when an omission fails to prevent a process from preventing another process from causing B, as in Figure 6.4(ii). For example, the mother, in failing to stop the father from grabbing the child, caused* the accident to not happen (i.e., prevented the accident).

The case described by Figure 6.4(ii) suggests one more interesting case. Suppose the counterfactual scenario of 6.4(ii) is actual. Then we have a case of causation* by the prevention of a prevention, as in Figure 6.4(iii). What is interesting about this case is that both A and B are real positive events. For example, the mother prevented the father from grabbing the child, thereby causing the accident. The details of the analysis need not detain us, but this is the case alluded to earlier where both relata of causation* are positive events. However, it involves the negative event not-*y* as an intermediate link.

Further, more complex cases are possible. Something, call it A, could prevent the kind of prevention-of-prevention shown in Figure 6.4(iii), thereby preventing B. Then we have a further case of A caused* not-B. This is why the analysis of preventions given in section 6.4 was presented just as a sufficient condition for prevention.

We also need to account for the prevention of possible causation, such as in 'the referee blew the final whistle, preventing any further

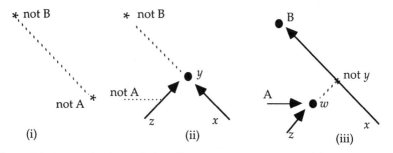

Figure 6.4. Prevention by omission (i) Type 1: not causing, and (ii) Type 2: failing to prevent a preventer. (iii) Prevention by prevention.

140

goals from being scored.' It is not the case that had the referee not blown the whistle, something would have brought about more goals. Rather, we need a might conditional in place of the would conditional of (P3) and (O3), roughly to the effect that 'had the referee not blown the whistle, someone might have scored further goals.'

There are also cases involving a kind of dependence other than causal, as in 'the father's staring up at the sky caused the accident.' Since the father failed to guard the child, by virtue of the fact that he was staring up at the sky, and the father's failure the guard the child caused* the accident, we want to say that the father's staring up at the sky caused* the accident. 'The father's failure to guard the child' depends in a certain kind of way, in the circumstances, on 'the father's staring up at the sky,' but not as cause and effect. So we need to allow not just for actual causal dependence as in the P and O definitions, but also for this other kind of dependence, in chains of causation*.

Finally, the present account has implications for the issues of delaying and hastening (Mackie 1992). According to Mackie, "typical delaying is bringing about by preventing whereas typical hastening is preventing by bringing about" (Mackie 1992: 493). If causation* and causation can themselves be relata of relations of causation* and causation, and supposing that if either relata of such a relation is causation*, then that relation is causation*; then delaying is causation*, since its cause* is a prevention; and hastening is causation*, since its effect* is an absence. The difference between the two is that delaying has causation* as its cause* and a positive as its effect*, and hastening has causation as its cause* and an absence as its effect*. I leave aside the question of whether this explains why we usually think that hastening causes whereas delaying does not.

Thus, as we saw in section 6.3, there is a lot more causation* in the world than first meets the eye. But this does not make everything a cause* of everything else. To have causation* there must be the possibility of genuine causation.

VI.7 SOME OBJECTIONS ANSWERED

In this section I will consider two objections to the above account, and show how I think they can be answered.

The first objection is that we could have a simpler account by analysing preventions as, roughly, A prevents B if A occurs and B does not, and if A had not occurred, B would have occurred. Similarly, omissions could be analysed as not-A causes B if A does not occur and B does, and if A had occurred, B would not have. Shouldn't this simpler counterfactual account be preferred, just because it is simpler?

Intuitively, it may seem that such a counterfactual theory might provide an extensionally adequate analysis of preventions. But this would miss the point. It may or may not be an extensionally adequate analysis of preventions, but as an answer to the problem set in this paper, it is inadequate, for the reason that it fails to explain the relationship between causation and causation*, for all theories of causation but the counterfactual theory. (Recall that in this account I am seeking to provide a cross-platform solution, that is, one available to any theory of causation.)

For Lewis's counterfactual theory, according to this alternative suggestion, causation* would not be causation because it involves negatives, which don't qualify as the kind of events that can be the relata of causation (see section 6.2). But yet causation* and causation both involve counterfactuals. So there is a difference, and there is also a similarity.

But for other platforms – that is, for non-counterfactual theories of causation – the relation between causation and causation* becomes mysterious. The alternative suggestion would make causation* different from genuine causation, but it would not show why we are justified in treating it as causation for practical purposes.[5] So the alternative suggestion does not provide a cross-platform solution account of prevention and omission.

The second objection is that my account of causation* entails that there is very much more causation* in the world than our intuitions allow. For example, it becomes possible that my not scratching my ear causes* a flood two years later in Mexico. However, this problem has its correlative in genuine causation: the Big Bang caused the flood in Mexico (see Lewis 1986, sensitive causation), which is a problem for any theory of causation that accepts indirect causation as causation. Rejecting indirect causation would wreak total havoc with causal talk; so too would ruling out 'indirect' preventings and omissions. So this is

---

5. A further problem is that the alternative suggestion is deterministic, in the sense that it makes A a necessary cause of not B (my theory doesn't, because causation itself might not be deterministic). However, this problem could be solved by a probabilistic version of the counterfactual theory of causation*, by analogy to Lewis's approach to chancey causation (Lewis 1986: 175–184).

not a special problem for causation*. Further, any solution to this problem with respect to causation can also be applied to answer the corresponding problem for causation*.

For example, my own view is that all indirect causes count as causes, including the Big Bang causing the flood in Mexico, and that if we want an account of the distinction our intuitions ask for between some cases of indirect causation and others, then pragmatic considerations need to be invoked. The same will be true of causation*.

VI.8 A CROSS-PLATFORM SOLUTION

In the definitions given here genuine causation remains primitive, which means that (to continue the computer metaphor) this solution should 'just work' on any platform, that is, for any theory of causation. That this is so can be seen by a quick survey of the theories and problems discussed in section 6.2.

The problem for the Conserved Quantity theory is that causation* lacks a set of causal process and interactions involving conserved quantities. This is solved simply by the fact that in the counterfactual about causation we will find the appropriate set of causal processes and interactions. The father's grabbing the child is a causal interaction, the child's running onto the road is a genuine causal process, and the accident is a genuine causal interaction.

Lewis's problem of having negative events as causes is also solved. The dilemma that Lewis faces is, we recall: take omissions as genuine events qua omissions and they are too disjunctive, take them as references to actual positive events and the appropriate counterfactual dependence does not hold between them and their effects*. With my account of omissions both horns of this dilemma are avoided, since we deny that omissions (and all negative events) are real events of the kind that can be the relata of causation. Then the counterfactual 'if not-A were not to occur, B would not occur' may well hold between omissions and their effects*, but on Lewis's theory of causation that won't mean there is genuine causation between them, because the cause is not the right kind of event.

So we avoid both of Lewis's problems. First, we are not allowing highly disjunctive events as causal relata, so we avoid the problems that brings. Second, we are not taking omissions to be the actual positive events that happen. If we did, there would be trouble. For my account of the omission 'not-A caused* B,' the trouble would be that the counterfactual 'if A had happened, A would have caused y' (where y is the interruption of the process that actually caused B) is not true, since if the father hadn't been looking at the flowers he might have been looking at the clouds. If we reject my account, we don't thereby solve this problem, and in any case Lewis is back with the problem that the omission and its effect are not linked by counterfactual dependence.

The difference for Lewis between causation and causation* would then be that causation (involving only positives) is the relation of counterfactual dependence, while causation* (involving negatives) is, taking the case of prevention, essentially the nested counterfactual 'had A not occurred, then had x not occurred, B would not have.' This allows Lewis to say that claims about causation by omission are true, but that there is a difference between causation and causation* that explains the difference intuition, and that there is sufficient similarity to warrant our treating causation* as causation for practical purposes.

Lewis could still say that omissions occur in some sense. 'The father failed to grab the child' can undoubtedly be a true sentence, made so by the actual events that happened; events such as 'the father was looking at the flowers with his hands in his pockets at the time the child was running towards the road.' (Lewis says that a higher-order fact to the effect that all the events that occur *are* all the events that occur is not necessary.) So omissions are some kind of event, but not the kind that genuine causation relates. After all, Lewis's account of events was written for his theory of causation (1986: 190). So, as was noted earlier, I am not ridding the world of negative events, just ridding causation of them (which is where they cause the most trouble). It's each to his own regarding how to analyse negative events themselves. (Note again that I am not claiming that my account is the only way to solve the problem. Indeed, in the previous section we noted an alternative solution for the counterfactual theory, although that solution is not a cross-platform solution.)

For Armstrong, the solution comes even more easily. Armstrong's problem was that causation is a second-order relation between first-order actual states of affairs, yet the truthmakers for negative 'states of affairs' (truths) include higher-order facts of negation. But prevention and omission seem to be causation involving negative states of affairs. On my account preventions and omissions are not real cases of causation, so the thesis that causation relates first-order states of affairs is saved. Causation* is not a genuine causal relation, but statements about it can be true; when they are, their truthmakers include negative 'states of affairs,' which are analysed by Armstrong in terms of certain relevant actual states of affairs together with some totality fact.

On the other side of the fence are those theories that do not recognise a real distinction between negative and positive events, such as Suppes's probabilistic theory. For them, causation and prevention are different names for the same thing, so they will deny that there is a real distinction between causation* and causation. However, if the distinction is accepted, then the theory will work: just plug in probability-raising relations for 'causation' in the definitions. This solves the problem that these theories face, viz., that they can't explain the intuition of difference.

It is true that negatives (negative facts or events) can be ends and means, and can raise chances. In fact, negatives, when they figure in causation*, play many of the roles that genuine causes and effects play. As well as serving as means and ends, since they raise chances they can be evidence for their effects* and causes*, and they can also feature in explanation. This helps show further why it does not matter that for practical purposes we don't bother to distinguish causation* from causation.

To summarise, the problems that negatives cause for a range of theories of causation can be solved by adopting the counterfactual theory of causation*: causation* is not genuine causation but a counterfactual truth about causation, in certain circumstances. Further, the intuition of difference and the theoretical problem of negatives in causation suggest that causation* differs from causation, and the counterfactual theory of causation* explains how they differ. On the other hand, the similarity between causation and causation* is explained by the unity between the two concepts: one is a counterfactual truth about the other. This similarity justifies our habit of treating causation* as causation, a habit to be expected given the epistemic problem of distinguishing them. That habit is also justified by the fact that causation* plays the same role as causation in evidence, explanation, and agency.

7

Connecting Causes and Effects

The account of causal processes and interactions – the Conserved Quantity theory – offered in Chapter 5 was cast as a minimal account in the sense that it does no more than explicate the key notions of causal processes and interactions. We may still ask how this is relevant to the events we call causes and effects. If two events (or facts, states of affairs, whatever) are related as cause and effect, what causal processes and interactions must there be? This second question, *what is the connection between causes and effects?*, is the subject of the present chapter.

We might expect that the process theory would provide necessary and sufficient conditions for singular causation in the following way: two token events are connected in a causal relation if and only if a continuous line of causal processes and interactions can be traced between them. Call this the naïve process theory. In section 7.1 we shall see that because of the problem of 'misconnections' this is inadequate as an analysis, the implication being that the account given in Chapter 5 needs to be developed or supplemented in some way.

In Chapter 2 we saw probabilistic (chance-raising) theories have problems with certain chance-lowering cases. A common way to deal with that problem is to supplement the chance-raising theory with some notion of a causal process. This suggests the idea that such a combination may also solve the problem of misconnections. Section 7.2 is a critical review of so-called 'integrating' theories – theories that attempt to provide a satisfactory analysis by combining the probabilistic theory with the process theory. I show that a number of extant integrating theories – due to Eells, Lewis and Menzies – are extensionally inadequate, and in section 7.3 I offer a new integrating account that avoids the problems facing the pure process theory and the chance-lowering accounts. The key is to recognise that chance-lowering causes are an example of what I call 'mixed processes.' However, it turns out that this too is unsatisfactory.

We return then to a pure process account, and drawing on Armstrong's metaphysics of states of affairs, I offer an account of causal relata, which together with an account of how they are linked by causal processes and interactions, solves the problem of misconnections. This enables us to see how causes and effects are connected together in a nexus of causal processes and interactions.

VII.1 THE NAÏVE PROCESS THEORY AND THE PROBLEM OF MISCONNECTIONS

What is the relation between a cause and its effect? A defender of anything like the Conserved Quantity account or the mark transmission account of causal processes and interactions, who seeks to discover the connection between causes and effects, might be tempted to answer along the lines of the naïve process account: two events (or whatever the causal relata are) are connected in a causal relation if and only if a continuous line of causal processes and interactions obtains between them.[1] For example, it is true that a shot in Sarajevo was the cause of the First World War just if there is a set of causal processes and interactions linking the shot with the minds of those various decision makers around Europe whose various decisions produced what we call the First World War. We shall examine both parts of this biconditional in turn.

1. *As a Necessary Condition.* In Chapter 2 we discussed a counterexample to probabilistic theories of causation, the case where a cause lowers the probability of its effect. The conclusion was that probabilistic theories do not provide a necessary condition for causation. The naïve process theory provides a ready answer to those counterexamples. Take the squirrel case, where a golf ball is travelling towards the hole, but a squirrel kicks the ball away, yet (improbably) the ball hits a tree and rolls into the hole.[2] There is a continuous line of causal processes and interactions that can be traced from the squirrel's kick to the ball's landing in the hole. The kick is a genuine causal interaction, and so is the collision with the tree. The movement of the ball is a genuine causal process. Thus there is, in Salmon's terms, causal influence produced by a causal interaction and transmitted by causal processes.

---

1. We should note that Salmon himself never explicitly claims that his theory does or should provide a sufficient condition. See Dowe (1992c: 196).
2. See Chapter 2, section 2.5.

The same may be said for the decay case, also discussed in Chapter 2, where the decay proceeds by an unlikely path, such that the cause lowers the chance of the effect (see Figure 2.2). An atom is a genuine causal process, and its decay is a genuine causal interaction involving, as it happens, not only production of the atom e, but also either an alpha particle or a beta particle. The decay has the form of a 'Y interaction.'[3] For the naïve process theory it is not of crucial importance that the cause raises the probability of the effect, but rather that there is a set of continuous causal processes and interactions linking the two events. Thus the naïve process theory accounts easily for these counterexamples. In fact, this has been one of the major arguments in favour of the process theory of causality (for example, Dowe 1993b; Salmon 1984: chap. 7). This gives us reason to suppose that the naïve process theory does provide a necessary condition for singular causation.

2. *As a Sufficient Condition.* However, the naïve process theory is not so successful in providing a sufficient condition. There are a number of cases that serve as counterexamples. One such case is Cartwright's sprayed plant: a healthy plant is sprayed with a defoliant that kills nine out of ten plants, but this particular plant survives (Cartwright 1983: chap. 1). We can provide a set of causal processes and interactions linking the spraying and the surviving, yet spraying does *not* cause the plant's survival. Papineau has given another example: being a fat child does not cause one to become a thin adult, although causal processes link the two. As in the sprayed plant case, two events, which we would not call cause and effect, are linked by a set of causal processes and interactions. According to Papineau, these counterexamples show that the naïve process theory of causality is inadequate (1989; 1986). Call this the problem of misconnections.

In fact, it can be shown that the failure of the naïve process theory at this point is *more* widespread and general than has been recognised. These counterexamples are not esoteric quibbles but a commonplace feature of causation.

Consider a tennis ball bouncing off a brick wall, a paradigm case for the process theory, since it involves clear-cut cases of causal processes and their interaction. The passage of the ball through spacetime is a causal process, and the collision with the wall is a causal interaction.

3. See Chapter 5, section 5.1; also Salmon (1984: 202) and Dowe (1992c: 212).

Now, certainly its hitting the wall is the cause of its rebounding. But the collision with the tennis ball does not cause the wall to remain in the same place, nor does it cause the wall to remain standing. Nor does the collision cause the ball to be green and furry. In fact, for any causal schema involving genuine causal processes and interactions there will be numerous events, facts, or states of affairs that are part of the schema. There is no guarantee that any two such events, facts, or states of affairs will stand in a causal relation: in general they will not, although some will. Thus spraying the plant with defoliant does not cause it to survive, or to still have mostly green leaves, or to still be in the corner of the yard; while spraying does cause it to be less healthy and to have some slightly yellowed leaves (let's say).

Similarly, the production of atom c was causally relevant to the production of atom e, but the decay's time (at dusk) and place (in a squash court) were not causally relevant to the occurrence of E. In each example, all of these events, facts, or states of affairs are part of a single schema of causal processes and interactions, yet many are not causally relevant. Indeed, these considerations raise the suspicion that the naïve process theory fails to provide a sufficient condition for *every* actual schema of processes and interactions. This applies to both the mark transmission version and the CQ version, for the events in question are linked by causal processes that transmit marks, and that display conserved quantities.

There are other kinds of misconnections as well. Take any two causally independent events, say, my tapping the table and the clock hand moving a moment later. There is no causal connection between these two particular events, but there is in fact a set of causal processes and interactions between them, according to the Conserved Quantity theory. The reason is that there are air molecules filling the gap between them, so that one can connect the two events by stringing together a series of molecular collisions. Also, there is a longitudinal disturbance, the sound of my tapping, which reaches the clock hand before it moves. So, again, there can be a set of causal processes and interactions where there is no causal connection.

Thus we must conclude that the naïve process theory as given does not provide a sufficient condition for singular causation; and hence that the naïve process theory is not an adequate account of the way causes and effects are connected. So how can a proponent of the process approach respond to this difficulty? This chapter addresses this question.

The conclusion reached in Chapter 2 was that the existence of a probability relation, be it PSR (positive statistical relevance) or counterfactual probabilistic dependence, between two events is not a necessary condition for singular causation between those events.[4] For example, a golf ball is heading towards the hole, and is kicked away by a squirrel, but, improbably, the ball rebounds off a tree into the hole. Further, if the argument of the previous section is right, it is a necessary but not sufficient condition for singular causation that the two events be linked by a causal process. So it seems that neither approach is adequate by itself. Thus, one line that has appealed to a number of authors is to combine these two insights into an integrated account of singular causation. Any such account I shall call an 'integrating solution.'

Of course, we cannot simply conjoin or disjoin the two notions. For an unnecessary sufficient condition conjoined with a necessary insufficient condition does not amount to a necessary condition; neither does an unnecessary sufficient condition disjoined with an insufficient necessary condition amount to a sufficient condition.[5] But perhaps there is some way to combine the two insights in some more subtle way so as to come closer to providing a full analysis.

Some authors have pointed to an asymmetry between the case of the squirrel and the case of the sprayed plant. The asymmetry is that while both have the same probabilistic structure, with the particular events occurring contrary to the governing statistical relations, nevertheless the squirrel's kick is a cause and the spraying of the plant is not. The same asymmetry exists between the decay case (similar to the squirrel) and the tennis ball case (similar to the sprayed plant). Some authors have used an explanation of this asymmetry as the basis for an account that brings together the PSR insight and the process insight.

Sober (1985) was the first to discuss this asymmetry, and, while noting that singular causation appears most 'opaque,' does make the suggestion that the asymmetry might be due to the fact that the squirrel's kick "introduces a new process" whereas the spraying does not

4. See Chapter 2, section 2.5.
5. Supposing that the probabilistic account does provide a sufficient condition, which is a dubious assumption (see Menzies 1989a; 1996).

(1985: 408). But, strictly speaking, this is not true. The spraying *does* start a new process, introducing certain active chemicals into the plant's physiology, while in a sense the squirrel's kick does not; it merely diverts the ball. In any case the claim that the spraying does not introduce a new process 'that caused' (1985: 408) or 'that led to' (Menzies 1989a: 648) the plant's survival merely begs the question, in the sense that it assumes that we already know whether one thing causes another, which is what the theory is supposed to tell us. So that suggestion won't work. Another possible reply, that the process introduced does not remain until the time of the survival, will not work either since the new process that the squirrel introduces (that is, a new direction for the golf ball) also does not survive until the sinking of the ball.

In this section I examine a number of theories that attempt to do this. I argue that each fails, but then in a later section I combine a suggestion of David Lewis's with the process account to provide an integrating theory that I claim can provide an extensionally adequate analysis.

VII.2.1 Probability Trajectory

In an extensive treatment in his book *Probabilistic Causality* (1991: chap. 6),[6] Ellery Eells develops a different proposal. He suggests that the asymmetry between cases like the squirrel's kick and the sprayed plant is due to different ways that the probability of an effect evolves between the occurrence of the two events. With this notion of continuous evolution of probability, Eells tries to draw together the probabilistic and process theories of singular causation. Eells gives definitions of 'because,' 'despite' and 'independent.' These definitions are explicitly offered as sufficient conditions, but we will be interested also in whether the theory provides a necessary condition:

An event E happens *because* of event C if (1) the probability of E changed at the time of C, (2) just after the time of C the probability is high, (3) this probability is higher than it was just before C, and (4) this probability remains at its high value until the time of E. (Eells 1988: 119)

Eells's definition of 'because' can be illustrated by a variation on the squirrel case. A player has played a good shot onto the green, and is poised to score a birdie until the ball is kicked by a squirrel at time t_C,

6. See also Eells (1988).

151

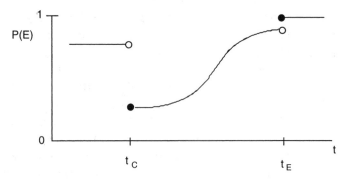

Figure 7.1. The sprayed plant.

after which the ball travels along an alternative route directly into the cup at time t_E. (Imagine that the alternative route follows a curved path around a slight incline to the side of the green, such that once on this path it is actually more likely that the ball will sink that it was when the ball was heading more or less directly towards the hole.) Condition 1 is met, as the squirrel's kick changed the chance of a birdie, that is, $P(E)$ changed at t_C. Condition 2 is met, since after the kick the birdie is quite probable, that is, $P(E)$ is high just after t_C. Condition 3 is met, since the birdie is more likely after the kick than before, that is, $P(E)$ is greater just after t_C than it is just before. Condition 4 is met, because the chance of a birdie stays high until it happens, that is, $P(E)$ stays high until t_E.

Negative causes are defined in terms of 'despite':

[A]n event E happens *despite* the event C if (1) the probability of E changed at the time of C, (2) just after the time of C the probability of E is low, and (3) this probability is lower than it was just before C. (Eells 1988: 120)

The paradigm example of a negative cause is the sprayed plant (see Figure 7.1). In this case the probability of the plant surviving until time t_E decreases to a low value just after time t_C, and then slowly rises later.

Third, event E is independent of event C if the probability is the same before and after t_C.[7] Thus independence means that neither

---

7. Eells says (1988: 119), "if the probability of E does not change around the time of C." It is not clear whether this means that the probability is the same before and after t_C, or that the probability does not change *at* t_C, or both. However, in his book Eells clarifies this as meaning just that the probability is the same before as after t_C (1991: 296).

condition 3 for 'because' nor condition 3 for 'despite' is met. Eells also defines an 'autonomous' event as one where the probability changes at t_C, the probability just after t_C is high and higher than just before, but the probability does not remain high, but drops down later (1991: 296). This is a case where the first three conditions in the definition of 'because' are met, but the fourth condition is not.

Thus the gist of Eells's theory is that the key probability is the probability *just after* t_C, rather than *at* t_C. The probability trajectories of the sprayed plant and squirrel cases differ in that in the squirrel case P(E) returns to its original high value straight after C, whereas in the plant case P(E) remains low for a substantial time after C as the plant slowly recovers. Thus the squirrel's kick causes the ball to sink, but the poisoning doesn't cause the plant's survival.

As Eells admits, this theory is not comprehensive, in the sense that there are many options not covered.[8] A more general point of criticism is that, although it does accurately describe a difference between the two cases (the squirrel and the sprayed plant), this theory is not based on any plausible intuition that this difference is the basis for a distinction between 'cause' and 'prevent.' Eells claims that its motivation comes from the way the theory naturally arises from the diagnosis of the standard examples (Eells 1991: 298). But when the extensional adequacy of the theory is examined, this view will be shown to be illusory. We now turn to that task.

It is easy to show that Eells's definition of 'because' does not provide necessary conditions for a singular cause. Condition 3 says P(E) must be higher after t_c than before. But, while this helps with cases like the squirrel and the golf ball, it does not solve the decay counterexample. Recall that the decay case is a counterexample to the thesis that causes raise the probability of their effects.[9] Event C, the decay of atom a into atom c, lowers the likelihood of event E, the production of atom e, since atom a will decay to atom b if it does not decay to c, and atom b will certainly decay into e. The particular instance where the decay process

8. For example, a case where an event raises the probability of a later event from a low value to a value of about 1/2 – for example, if drinking a bottle of kerosene kills with probability 1/2 and Adam drinks a bottle of kerosene and dies. Other cases not accounted for include events that lower the probability to a high or middling value, or raise it to a low value, or where the probability does not change at t_C, but does so immediately after.
9. See Chapter 2, section 2.5.

moves from A → C → E gives a trajectory for P(E) that drops at C *and remains lower*, yet we call C a cause of E. More specifically, the probability trajectory develops as follows: up until and immediately before C, the probability of E is 15/16, and from C until immediately before E, the probability of E is 3/4. This fails condition 3 of Eells's definition, which requires that the probability of E immediately after C be higher than it was immediately before C.

There are counterexamples to the other conditions as well. Condition 1 requires that P(E) change at t_c. A counterexample to this is the case where the ball is rolling along the green following a poor shot, and is not likely to drop into the hole. A squirrel kicks the ball, an event that does not really change the probability of a birdie. However, improbably enough, the ball takes a route favourable to sinking, and then in fact does sink. This is perfectly plausible: if a token cause can lower the probability of its effect, then most surely it can leave it unchanged. Thus the probability does not change at t_c, and so condition 1 is not a necessary condition.

Condition 2 says that P(E) must be high just after t_c. A counterexample to this is also easy to find. Suppose smoking as a youth can cause cancer as an adult, with probability p = 0.1. Consider a case where Adam smokes as a youth, then gives up smoking, but gets cancer as an adult. Then we have a case where the probability of the effect is not raised to a high value by the cause. This will happen for the entire class of low probability causes. In fact, it is quite unclear why Eells wants to have this condition, as it would not seem to cause any problems for his analysis if it were dropped.

Condition 4 says that P(E) must remain high until t_E. A counterexample to this is also easy to find. Suppose the ball has been deviated by the squirrel along the alternate route. But a high wind (improbably) springs up, which then makes it considerably less likely that the ball will sink. But, improbably, the ball does go in, despite the wind. Then the kick still causes the birdie, since the wind fails to prevent the birdie, and the kick remains the basic cause. According to Eells, this is a case of 'autonomy,' which is the wrong result: it ought to be a case of 'because.' So condition 4 is not a necessary condition either.

Thus it is clear that Eells's theory does not provide a necessary condition for singular causation. In fact none of the conditions Eells lists are necessary conditions.

Eells's theory does not even supply a sufficient condition, which is what it explicitly claims to do. For example, consider a case where the squirrel's kick sets the ball off in a path further away from the hole. But immediately after the kick, a strong wind springs up and blows the ball towards the hole, and into the hole. Then the kick is not a cause of the birdie at all, yet it qualifies as one according to Eells's theory. Thus we conclude that Eells's theory fails to provide a sufficient condition for singular causation. It seems that this approach, where the shape of the probability evolution is the key, does not work. It inherits the defects of the probabilistic account, and has further troubles of its own making.

VII.2.2 Chains of Probabilistic Dependence

We shall now examine two suggestions that attempt to integrate the probabilistic and process insights via the notion of a chain of probabilistic links. David Lewis (1986: 175–184), recognising that his probabilistic dependence does not provide a necessary condition for causation, uses that notion to define a 'chain of probabilistic dependences' (in Menzies' wording [1989a: 650]). A chain of probabilistic dependences is an ordered sequence of events such that each event probabilistically depends on the previous event. Call this a Lewis chain. Then according to Lewis, *C is a cause of E if and only if there is a Lewis chain between C and E.*

Peter Menzies has shown that the Lewis chain fails to provide a sufficient condition for singular causation. Consider the case where a reliable chain of neurons a-c-d-e is preempted by an unreliable chain of neurons b-f-g-e. Neurons a and b fire at the same time, but neuron c is inhibited by neuron b, which is 'moderately probable' (Menzies 1989a: 646). (That the inhibitory firing be 'moderately probable' is not necessary in this counterexample; Menzies sets it up this way for a different reason.) Take the events A through E to be the events that neurons a through e fire. $P(E|A) > P(E)$, and if a were to fire, e would be more likely to fire than it would be if a were not to fire. According to Lewis's definition, firing of b causes the firing of e, because there is a Lewis chain B-F-G-E. But, Menzies points out, A is also a cause, because there is a chain of probabilistic dependences A-E, where if A were not to occur, E would be much less likely to occur (1989a). But A is not the cause of E, as it happens; therefore the Lewis chain is not a sufficient condition. This sort of problem will arise whenever what is likely to cause some effect is preempted by an event less likely to cause that effect.[10]

In fact, it is easy to concoct more general examples. Suppose there is no inhibitory link between neuron b and neuron c, but that otherwise the mechanisms are identical to those in the previous case. And suppose that in a particular case a and b fire simultaneously, and that c, improbably, does not fire, but f and g do, causing e to fire. Then, on Lewis's theory, A causes E, because E is more likely to occur if A were to occur than if A were not to occur. Such counterexamples will arise whenever a probable cause fails to act, yet some other less probable mechanism produces the same effect.

Menzies offers his own version of the chain, which also is an attempt to incorporate both the probabilistic and the process insights into an account of singular causation (1989a). For the probabilistic part, Menzies uses Lewis's notion of probabilistic dependence, and for the process part, Menzies introduces the notion of an 'unbroken causal process.' There is an *unbroken causal process* between events C and E if and only if for any finite sequence of times between C and E, there is a corresponding sequence of events that constitutes a Lewis chain (of probabilistic dependences). Menzies uses this notion in his definition of cause: *C is a cause of E if and only if there is a chain of unbroken causal processes between C and E*. Call this the Menzies chain. The effect of this is to allow one to cut the chain at convenient places.

Before we consider the adequacy of this theory, we will need to show how Menzies can interpolate events not already given. This is important because a causal process requires that for any time between the cause and the effect, there is a corresponding event that stands in the appropriate relation to the cause and to the effect.

Suppose neuron c may fire two axons, one of which stimulates a certain neuron e, while the other fires it.[11] Take a case where e is stimulated. We want to know if there is a chain of unbroken causal processes linking C and E. For this to be so, for any time t_i ($t_c < t_i < t_e$)

10. Ned Hall offers a solution to this problem: take the chance just before the effect occurs. Then A's occurrence doesn't change the chance of E, since C has already failed to occur (Hall unpublished: note 6).
11. If this is not actually how neuron systems operate, then consider Hesslow's famous case of thrombosis caused by pregnancy and by contraceptive pills (1976).

there must be an event I which is more likely to occur if C has, and which makes E more likely. This requires that we can interpolate events at any time. According to Menzies's account of the causal relata (1989b), we need to identify 'real situations,' which are constituted by a property, an object and a time. Here we may simply identify the axon at t_i and the appropriate properties that make the firing likely. In this case event I will stand in the appropriate relations to C and E for any t_i. Hence C causes E.

Menzies's theory is clearly superior to Lewis's in the sense that it solves cases where the cause of some effect is preempted by an event less commonly the cause of that effect. Recall the difficulty for the Lewis chain: a reliable chain of neurons (a-e) is preempted by an unreliable chain (b-e). According to Lewis's definition, a is a cause because there is a chain of probabilistic dependences a-e. But according to Menzies's definition, b causes e because there is an unbroken chain of processes b-f-g-e, and a is not a cause of e because there is no unbroken chain of processes between a and e. The key to this success is that it requires a spatiotemporally continuous chain.

Menzies correctly claims that his theory will also work for other cases considered previously. In the case of the squirrel's kick, it is possible to trace out a chain of unbroken causal processes between the kick and the ball's sinking. In the case of the sprayed plant, however, it is not. The process begun by the poison is cut short by other processes (1989a: 656–657). So we may surmise from this that Menzies's theory is able to provide a sufficient condition.

When we turn to the case of the decay schema, and the question as to whether the chain theory provides a necessary condition for causation, we find first that it *does* handle the decay case as given in Chapter 2, but that it does *not* handle a variation on that case. This is so for both the Lewis chain and the Menzies chain. Recall again that C (the production of atom c) is causally relevant to E (the production of atom e), yet its occurrence lowers the chance of E. Recall also that Lewis's theory of probabilistic dependence fails to account for this case, because E would be more likely if C were not to occur than if C were to occur.[12]

But the Lewis chain enables us to interpolate an event C′, the existence of atom c at a time between the occurrences of C and E. There is a relation of probabilistic dependence between C and C′, because if C were not to occur then C′ would not occur, except in the unlikely

12. See Chapter 2, section 2.5.

scenario that a c-atom is produced by a process other than the decay of that a-atom. Thus if C were not to occur, C′ would be very much less likely than if C were to occur. Also, there is a relation of probabilistic dependence between C′ and E, because if C′ were not to occur then E would almost certainly not occur. This follows from the Lewis approach to interpreting these counterfactuals: in supposing C′ not to occur, we hold fixed the world up until the time of C′, which means that C occurred (and so B will not), and that the atom c has disappeared; so there's virtually no chance that E will occur. Thus there is a Lewis chain linking the cause and the effect, comprising C-C′-E.

There is also a Menzies chain. For any time t_i between the times of C and E one can define the event C_i, the existence of the c-atom at time t_i, such that C-C_i-E forms a Menzies chain. This will work for any number of events of this kind at times between the times of C and E. Thus on either account – Lewis's or Menzies's – the production of c counts as the cause of E, even though it lowers its chance.

However, these theories are not successful with a simple variation on the decay case. In this hypothetical variation, we have the same probabilistic structure: $P(C|A) = 1/4$, $P(E|C) = 3/4$; and the particular decay moves from A-C-E. But in this variation we have a *cascade*, where the c-atom immediately decays to e. Further, in this variation we suppose that time is discrete, and that the decay of c to e occurs at the very next instant following the decay of a to c. Then we can think of the second decay as occurring instantaneously. Thus there is no time between the times of C and E, and so there is no event C′; and since C and E do not stand in the relation of probabilistic dependence, there is no Lewis chain between the cause and the effect. And by similar reasoning, there is no Menzies chain between the cause and the effect. Therefore these accounts do not supply a necessary condition for singular causation.

The following reply fails. Suppose we define E* to be the event 'atom c decays by α decay.' Then take the sequence C-E*-E. This qualifies as a Lewis chain, one might argue, because if C were to occur, the chance of E* would be greater than if C were not to occur (where the chance of E* is zero); and the chance that E would occur is greater if E* were to occur (it would be one) than if it weren't. So by this theory, C is a cause of E.

However, I do not think this reply works. The mistake is to take E* and E as different events. They are not. They are physically equivalent, two different descriptions of the same event – the specific mode of decay in operation in this instance. The description 'α decay' refers to the breaking of c into α and e. The only difference is that E* is a more finely grained description. Thus E* is not a separate event, and so not a suitable linking event. So C-E*-E is not a Lewis chain. If, on the other hand, we take C** to be the appearance of the a particle, then C** and E are not events in a chain but concurrent effects of a common cause. So this reply fails.

Another type of counterexample to these chain theories is the case of deterministic overdetermination. Guns a and c both fire at glass e, and both are certain to break it, if it's not already broken. The bullet from c in fact hits the glass at the same time as does the bullet from a. So we have a case of deterministic overdetermination.[13] But, on Menzies' account, shooting the gun c is not a cause of the glass breaking, because there is no chain of unbroken causal processes linking c and e. To see this, consider the chain C-E. The effect E was not less likely to occur if C had not; either way, it is certain to happen. Or take an event I, the bullet from c travelling towards the glass at time t_i. The events I and E are not linked by an unbroken causal process, because E is just as likely to occur if I does not. Clearly, this will hold for I at any time t_i. If this fails on Menzies's account, then it will fail on Lewis's account, because a Lewis chain is a chain of probabilistic dependences, and as we've seen there's no event I in the trajectory of the bullet from a such that e probabilistically depends on I.[14]

So neither of these chain theories turns out to be an adequate integrating account.

VII.2.3 Quasi-dependence and the Canberra Plan

Two more promising suggestions that we shall consider are again due to Lewis and Menzies: Lewis's notion of 'quasi-dependence' (1986: 205–207), which is in effect an 'integrated account', and Menzies's recent functionalist-style account, discussed in Chapter 3 – which I will call the Canberra Plan.[15]

13. Compare what Lewis (1986: 193) calls 'redundant causation.'
14. Another problem is that, like Salmon's theory of mark transmission, Menzies chains must be spatiotemporally continuous. Since it may be that quantum processes involve discontinuities, this is undesirable. Menzies claims that such a possibility is "too far fetched" (1989a: 651), but since I think I can provide the analysis without ruling out discontinuities, I regard this as a further defect.
15. This is the name which has been applied to the use of a Lewis-style functionalist approach to various areas of philosophy as developed, for example, by members of the philosophy department at the Australian National University in Canberra. (The name was coined in O'Leary-Hawthorne and Price 1996.)

In response to cases where causes do not stand in the right relation of probabilistic dependence because of the existence of some external event (such as the decay case or the preempted neuron case), Lewis calls upon the intuition that only what is intrinsic to a process can be relevant to that process being causal. To cash out this intuition, Lewis considers a process that does not exhibit the appropriate probabilistic dependence because of some extrinsic factor, but where the great majority of similar processes in other regions of spacetime do. Lewis proposes that such processes be said to stand in the relation of *quasi-dependence*. Quasi-dependence may be defined as follows: if and only if two events, C and E, both part of the one process, where E does not probabilistically depend on C but in other spacetime regions where E-type events and C-type events are part of the one process, the E event normally probabilistically depends on the C event, then E quasi-depends on C. We may then redefine the Lewis chain as a sequence of two or more events with either probabilistic dependence *or* quasi-dependence at each step.[16]

But Lewis's account is brief, and he does not address possible difficulties. One question the account raises is: for Lewis, what is a process? For, in order to determine quasi-dependence, we must know what it is that is to be the same in other regions. We, if not Lewis, may answer in terms of the CQ definition: a process is the world line of any object to which may be ascribed a conserved quantity. In fact, the intuition that causation is intrinsic to a process is explained if it is true that causation is to be analysed according to the CQ theory.

Second, how similar may we allow these processes to be, and similar in what respect? This latter question proves to be quite a difficulty when we consider the capacity of this suggestion to provide a necessary condition for singular causation. We will return to that question shortly.

In a more recent paper (1996), Peter Menzies has offered a new account to replace his earlier theory (1989a). His reasons for abandoning the Menzies chain as the appropriate explication of singular causation are different from the main objection developed in the previous section. His first two reasons concern the connection between the Menzies chain and our folk intuitions about causation. He now believes

---

16. The quasi-dependence theory unfortunately does not solve the problem raised by Menzies. The problem is that if an event A reliably leads to event E (as in Figure 2.2), E will both probabilistically depend and quasi-depend on A. So Lewis's modification wrongly counts A as the cause of E. But in this case A does not cause E, we would think. Hence the quasi-dependence suggestion, like the Lewis chain, does not provide a sufficient condition.

that it is too strong a conceptual requirement to rule out action at a temporal distance, and that the definition is too intricate to capture the ordinary folk notion. His third reason concerns the case of deterministic overdetermination, as given above.[17]

Menzies's new theory[18] takes the causal relation to be a theoretical entity, and adopts the style of analysis associated with Lewis's approach to the analysis of mind (see, for example, Lewis 1994). Menzies takes chance-raising to be one of the central tenets of the folk theory of causation, which, if true, allows the causal relation to be defined as the relation that *typically* raises chance. Menzies then suggests that a contingent identification may be made between what typically exists when there is chance-raising, and energy transfer. It would then follow that in our world causation is energy transfer.

Both Lewis's quasi-dependence and Menzies's Canberra Plan can deal with occasional improbable recalcitrant cases. However, this strategy fails in the decay case. Let's start with Lewis. When we define 'similarity' in the most natural way, it seems that the quasi-dependence theory does not provide a necessary condition. The most natural way to approach the decay case is to consider other spacetime regions where an atom of type c decays by α decay to produce an e atom. There, we would hope, C would raise the chance of E. However, it may not do so. The reason is that it is quite plausible that c-atoms most commonly form from a-atoms, and in that case C events will not normally raise the chance of E events. In other words, although we may say that atoms decaying by α decay almost always form Lewis chains, we may not say that in other spacetime regions the atom c's decay to e forms a Lewis chain, for it will not, supposing that c-atoms most commonly form from a-atoms. It's not just this instance of the decay pattern, but all such occurrences that will be problematic. It should not be difficult to find a real example of that phenomenon.

Another idea is to appeal to natural kinds in the sense that Lewis utilises in other contexts. But it's not clear that this would avoid the difficulty, since c-atoms are natural kinds. Other ways to cash out 'similarity' in the decay case are to consider 'all atoms decaying by β decay' and 'all atoms decaying.' In these cases the present difficulty may be overcome, although the strategy does seem arbitrary and ad hoc. We shall return to this point.

17. A point made previously in my paper (unpublished-b).
18. Discussed in detail in Chapter 3 of the present work.

Menzies claims for his new account that it handles without difficulty cases of probabilistic preemption. The part of his theory that permits this is the word '*typically*.' For this allows that there be *some* cases that exist where there is causation but not chance-raising,[19] so long as the cause usually raises the chance of the effect. Menzies takes the standard chance-lowering causes – discussed earlier – to be exactly this sort of case. This strategy is similar to the strategy adopted by Lewis: preempted causes normally raise the chance of their effects, in most parts of the universe, and it's just on this occasion that we happen to have another process interfering and adversely affecting the chances.

But Menzies's new theory fails for exactly the same reason that the quasi-dependence theory fails. The decay case is an example of a chance-lowering cause that does *not* typically display chance-raising. The interesting thing about that case is that the processes associated with it are typically associated with it, as that is the normal mode of decay for that atom. Indeed, the decay case has a *nomological* character to it, whereas most cases of chance-lowering causes in the literature are fortuitous de facto constructions. So Menzies's new approach does not avoid the difficult cases. It too fails to provide a necessary condition for singular causation.

Nevertheless, Menzies's account may avoid this criticism, in the sense that chance-raising may still serve as a reliable but fallible reference fixer for causation in everyday situations, even if it is not an adequate analysis of causation. I am happy to grant this (we will see in the next section in what kinds of situations it fails). However, as an account of what causation actually is, chance-raising fails, as my objections show. The objections apply especially to the account of Mellor (1995), in which chance-raising is the central connotation of causation.

VII.2.4 Other Accounts

Finally, there is one more class of 'integrating solutions' to consider. These attempt to deal with chance-lowering causation by conditionalising on events between the cause and the effect, or something similar (Kvart 1997).

According to Kvart, A causes B just if A has 'some positive causal impact' on B, which is the case just if there is a 'strict increaser' for A

19. And vice versa.

and B (assuming A is causally relevant to B). C is a strict increaser for A and B just if C is an increaser that is not 'reversed' by another event between A and B, where an increaser is an intermediate event that reverses the 'initial probabilistic indication.' A strict increaser is one for which there is no further reverser.

Kvart understands the probabilities in terms of a counterfactual analysis in which the world is held fixed until the time of the cause, and in addition, information between the cause and the effect is admitted, namely, the set of true factual statements to which the cause A is causally irrelevant or purely positively causally relevant (Kvart 1997: 407).

So in the case of the pulled drive p that causes the hole in one h but lowers its chance, the initial probabilistic indication is one of chance-lowering. However, we must

consider events along the untraversed route, in particular actual events featuring the non-occurrence of non-actual events that would have occurred had this route been traversed. (1997: 416)

The event g that the golf ball didn't travel on the path along the fairway is an increaser of p and h, since the chance of a hole in one given p and g is greater than without p but with g. But there are no other events that can reverse this result. So g is a strict increaser for p and h, and therefore p has some positive causal impact on h.

However, this account and others like it cannot deal with the decay case. The reason is simply that with cascading in discrete time there is no event between c and e, on either the traversed or the untraversed route.[20]

The integrating accounts considered in the previous section all veered closer to the chance-raising account than to the CQ theory. It is not surprising, therefore, that they handle the misconnections case but not the chance-lowering causes. This suggests that what we need is an integrating theory that takes the notion of a causal process more seriously, but that can handle the chance-lowering causes. The first step in this account is to recognise the difference between causal processes and another kind of process.

20. The same can be said for the Σ-set account of Paul Noordoff (1999).

163

There can be more than one process between a cause and its effect. Sometimes these processes are all causal processes: a gun fires a bullet that severs a rope, causing a large rock to land on a person's head just as the same bullet continues its path through that person's heart, a case of causal overdetermination. In other cases we have what might be called mixed processes: the processes between the cause and the effect are not all causal processes as defined by the CQ theory; some are processes of prevention.

We saw in the previous chapter that 'causation' by omission and prevention (causing not to occur) (and other kinds of causation involving negative facts or the nonoccurrence of events) are to be understood not as cases of genuine causation, strictly speaking, but in terms of the mere possibility of genuine causation. For example, 'the father's grabbing the child prevented the accident' is to be understood in terms of the possibility that an accident would have been caused by certain circumstances had the father failed to act. This means that only some of the processes linking the cause and the effect are genuine causal processes.

Chance-lowering decay is an example of mixed processes. In the decay case (Figure 7.2), in the particular instance where the process moves A-C-E, there is an actual causal process and a preventing process. Event C, the decay of atom a to atom c, also prevents another processes from going through, namely the process from A-D-E. On the strength of this, C is linked to E via a preventing process. The preventing process, which in fact fails to prevent the effect, goes from C to not B to E.

In general, I claim, chance-lowering causation arises where an event C initiates two processes, one of which – call it ρ – has a chance of causing E, the other – call it σ – a chance of preventing E, and where the actual causal process is more reliable than the prevented causal process. Causes that initiate mixed processes act at the same time to cause and to prevent the effect. Of course, they cannot successfully do both, although they can fail to do both.

This formulation might seem a little awkward in the decay case, since C does not, strictly speaking, initiate a process that might have prevented E. Rather, simply by occurring, it prevented the reliable process A-D-E from producing E. However, it is common practice in philosophy to speak of preventing A by bringing about something incompatible with A (e.g., Gasking 1955).

164

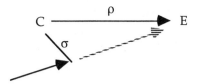

Figure 7.2. Decay case as involving "mixed paths."

Depending on the actual probabilities involved, there are two possible problems with mixed processes. First, if $P(E|C) < P(E)$ and process ρ is successful, then C causes E and we have a chance-lowering cause. Second, if $P(E|C) > P(E)$ and process σ is successful, then C prevents E and we have a chance-raising prevention. Chance-lowering causes occur whenever the former obtains.

To take a well-known example due to Hesslow (1976), contraceptive pills promote thrombosis as a side effect, but they also hinder it by virtue of the fact that pregnancy prevents thrombosis. The probabilities are such that taking the pill significantly lowers the chance of thrombosis. Suppose Mary takes the pill, does not get pregnant, but does get thrombosis. Then there are mixed processes between her taking the pill and her getting thrombosis, the weaker one a successful cause and the stronger one an unsuccessful 'preventer.'

VII.3.2 An Integrating Solution

Had there been just the one process – ρ, say – then it would have been the case that C caused E and raised its chance. Had there been just process σ, then it would have been the case that C prevented E and lowered its chance. The problems arise, we can now see, when an event C is both potentially a cause and potentially a preventer – cases of mixed processes.

I think there is a way to combine the process and the chance-raising insights so as to account for the mixed processes. To begin, consider Mellor's concept of the chance that C gives E in the circumstances, which he writes as $ch_C(E)$. In cases of multiple processes between a cause and its effect we can (although Mellor doesn't) think of that chance as having components. Just as the gravitational force on the moon has a component due to the earth and a component due to the sun, so the chance that C gives E has a component due to the possibility of process ρ and a component due to the possibility of process σ.

165

In the decay case (Figure 2.2), A gives E a chance of 15/16, which has as components the chance 3/16 that E will be produced via C, and the chance 3/4 that E will be produced via B. Following the terminology of Mellor (1995), we can write this as:

$$ch_{A\rho}(E) = 3/16$$
$$ch_{A\sigma}(E) = 3/4$$

where $ch_{A\rho}(E)$ reads 'the chance that A gives E in virtue of ρ.'

My suggestion is, then, that for C to be the cause of E by virtue of process ρ, then it must be the case that

(A) C would raise the chance of E were ρ the only process between C and E.

Whether C raises the chance of E at that closest 'ρ-only' world is itself a counterfactual matter, analysed as a closest world conditional (assuming here that we are following the counterfactual approach to chance-raising).

(A) is not a counterfactual that can be analysed in the usual manner of holding fixed the world until the time of the antecedent, because process ρ begins, temporally, with C. What we need to do is to compare the ch(E) at the closest worlds to ours where ρ is the only process involved between C and E, with the ch(E) at the closest worlds to that world where C does not occur. The ch(E) at the closest worlds to ours where ρ is the only process involved between C and E should be the component $ch_{C\rho}(E) = 5.0 \times 10^{-7}$, whereas the ch(E) at the closest worlds to that world where C does not occur is $ch_{-C\rho}(E)$, which in our decay example is zero, since C is a necessary cause of E. Thus

$$ch_{C\rho}(E) > ch_{-C\rho}(E),$$

so we say C is a cause of E by virtue of process ρ.

Similarly, for C to be the preventer of E by virtue of process σ, then C must lower the chance of E were σ the only process involved between C and E. In our case, $ch_{C\sigma}(E)$ is zero, while $ch_{-C\sigma}(E)$ is one, since if C doesn't occur D occurs, which leads to E with probability 1.

Note also that in the decay case the ρ-only world is counterlegal, supposing that such decay schemes constitute laws of nature, although in other cases, such as slicing a golf ball into the hole via a tree branch, the relevant ρ-only world is not counterlegal.

This approach succeeds where others fail. In certain cases of mixed processes we can conditionalise on different parts of C, if, for example, the atom's having one quantity is responsible for one process and its having another quantity is responsible for the other (in the manner of the 'more closely specification of events' solution to some putative chance-lowering causes – see Chapter 2 and Salmon 1984: chap. 7). But in general it may be the same quantity responsible for and involved in both processes. In our decay example this is indeed the case. Further, there is no way to conditionalise on unknown parts of C, because we already have all the information there is, if the indeterministic interpretation of quantum mechanics is correct.

We can also see why the accounts discussed in section 7.2 work to the extent that they do. In effect, conditionalising on an event between the cause and the effect takes you to the closest ρ-only world. But they fail if there is no such intermediate event.

So the integrating account of causation that I am proposing is as follows:

C causes E iff

1. there is a set of causal processes and interactions, as defined in Chapter 5, between C and E, and
2. $ch_{Cp}(E) > ch_{-Cp}(E)$, where ρ is an actual causal process linking C with E.

To summarize, this account solves the problem of the sprayed plant and other similar misconnection cases by introducing to the Conserved Quantity theory a chance-raising condition. But it avoids the problems with chance-lowering causes by distinguishing components of objective single-case chance, delineated according to relevant possible paths between the cause and the effect. These processes are themselves delineated by the original Conserved Quantity theory.

VII.3.3 Further Problems

However, while this account does, I believe, show how causes raise the chance of their effects, it is not satisfactory for two reasons. The first concerns further cases of misconnections. Suppose that particular event C occurs, and raises the chance of event E, but that it doesn't cause E, but that E occurs by chance or by some other cause. Suppose my tapping the table can cause the clock hand to move, because the clock is rigged up to a microphone, and suppose that there are other ways that the hand might move, for example, by its normal timing mechanism. Suppose I tap, and perchance the microphone fails to pick up the signal, but the clock ticks anyway. Then my tapping the table raised the chance of the hand moving, and the two are linked by a set of causal processes and interactions, since the sound wave does strike the clock hand. Then even by the integrating account of the previous section, it is a cause.

The second issue is that even if the analysis just presented does show how all causes raise the chance of their effects, that does not mean that we should necessarily analyse causation in terms of chance-raising. Perhaps we need to analyse chance in terms of causation. Although this theme is beyond the scope of this book, I think this is exactly how we should account for chance.

For these reasons I think that we need to abandon the integrating approach to analysing causation, and return to the pure process account to see if it can be developed so as to overcome the difficulties it faces without the help of the notion of chance. We begin this task in the next section.

VII.4 CONNECTING CAUSES AND EFFECTS

We want to address the question of what relevance the Conserved Quantity theory might have to the connection between the things that we think of as causes and effects. We saw in section 7.1 that the naïve process account, which says two events are connected in a causal relation if and only if a continuous line of causal processes and interactions (as defined, let's say, by the Conserved Quantity theory) obtains between them, fails for a large class of cases, which we called 'misconnections.' In this section we show what causes and effects are and how they are connected, and how this deals with the problem of misconnections. First, taking the cue from Armstrong's metaphysics of states of affairs, we outline an account of causal relata, and second, we show how these are connected by causal processes and interactions in such a way as to rule out misconnections.

VII.4.1 Armstrong's World of States of Affairs

According to David Armstrong (1997), the world is a world of states of affairs, that is, items such as a particular having a property or standing in a relation to another particular. States of affairs can be more complex; the constituents of molecular states of affairs may include 'atomic' states of affairs, if such there be. States of affairs are contingent existents, and there are no logical relations between first-order states of affairs.

However, there are first-class states of affairs, whose properties and relations are genuine universals (the best candidates for these are from science), and second-class states of affairs, whose properties and relations are not universals, and which obtain contingently (1997: 45). All the second-class states of affairs supervene on the first-class states of affairs, which in Armstrong's sense of 'supervene' means that there can be no difference in second-class states of affairs without a corresponding difference in first-class states of affairs. According to Armstrong's doctrine of the ontological free lunch, that which supervenes represents no increase in being.

Drawing on Sellar's distinction, Armstrong dubs the commonsense world of loose second-class states of affairs the 'manifest image,' and the first-class world of states of affairs the 'scientific image,' so that the manifest supervenes on the scientific. Epistemologically, our access to the scientific image is via the manifest, but Armstrong rejects the Wittgensteinian semantic descent from the second-class to the first-class.

There are no negative (as we saw in the last chapter) or disjunctive universals, so there are no negative or disjunctive first-class states of affairs. Even if 'Pa' (particular a has property P) is a first-class state of affairs, then although it is logically equivalent, 'Pa.Qa v Pa.~Qa' is not a first-class state of affairs.

Causation is singular, a contingent higher-order relation between first-order states of affairs. Conceptually, causation is primitive, but ontologically, it is an instantiation of a law of nature (contingent connections between state-of-affairs types) (see also Armstrong 1999).

In what follows we will draw on Armstrong's account of the relata rather than the relation of causation, with some differences in terminology. I set aside the question of the relation between the Conserved Quantity theory and Armstrong's account of causation.

VII.4.2 The Relata of Causation

We shall suppose that the causal relata are either events or facts, both of which concern objects having properties at a time or a time period. An *event* is a change in a property of an object at a time, for example, a quantitative change; or a related simultaneous change in more than one property of more than one object at a time, and so on. (This is a thin sense of 'event'; we leave it open how a thick sense may be worked into the Conserved Quantity theory). A *fact* is an object having a property at a time or over a time period. Because both events and facts concern objects, this fits well with the Conserved Quantity theory. For simplicity we will deal just with facts.

Furthermore, we will take it that such facts or events, if they enter into causation, must involve conserved quantities or supervene on facts and events involving conserved quantities. For example, the fact that the ball is green must supervene on the fact that various bits of the surface of the ball have certain physical properties by virtue of which the ball looks green. If these properties are not conserved quantities, then they in turn must supervene on conserved quantities. This seems to be a natural development of the Conserved Quantity theory. Then we can write the relevant fact as $q(a) = x$, which reads 'object a has x amount of conserved quantity q' or, to abbreviate, simply 'q(a).' If a second type of conserved quantity is involved, we will write this as 'q′.'

We will also distinguish between the manifest and the physical (the latter in order to be in keeping with the terminology of this book). So we say that manifest causal facts supervene on physical causal facts.

No physical causal facts are disjunctive or negative. This is important for understanding prevention and omission, as we saw in the last chapter, and for understanding misconnections, as we will see in the next section. It is important to note that this rules out facts like 'having an amount less than x of q.' For example 'the temperature is less than Y' appears to be a disjunctive fact, equivalent to 'the temperature is absolute zero or 1 or 2 or . . . or X or . . . or Y − 1,' which, if the temperature is in fact X, is true by virtue of the fact that X is one of the disjuncts. So facts involving 'is less than' are not physical causal facts. (Note that the Armstrong supervenience thesis does not assert that everything that supervenes on a genuine physical causal fact is a causal fact.)

VII.4.3 The Causal Connection

As we have seen, not every set of causal processes and interactions constitutes a genuine causal connection. Which ones do?

Suppose I_1, I_2 are causal interactions connected by causal process p.

I_1 and I_2 are defined in terms of the exchange of certain conserved quantities, and p is defined in terms of the possession of a conserved quantity. My suggestion is that we therefore characterise 'causal connection' roughly as follows (a more precise version will be given later):

Causal Connection: Interactions I_1, I_2 are linked by a causal connection by virtue of causal process p only if some conserved quantity exchanged in I_2 is also exchanged in I_1, and possessed by p.

Further, I suggest that the things we call causes and effects (facts, events, or whatever they are) at base involve the possession of, or change in the value of, some conserved quantity. For simplicity, think of causes and effects as causal interactions. Then we can say that causes and effects are linked by a causal connection by virtue of the fact that a quantity exchanged in the 'cause' is possessed by a causal process that links the cause to the effect, where it is also exchanged.

What about more complex cases, where different properties are exchanged 'in the cause' as compared to the effect? The answer is, I think, that if we think of the link between the cause and the effect as involving a string of causal processes and interactions, then somewhere along the line the quantity exchanged in the cause and the quantity exchanged in the effect must together be exchanged in the one interaction.

To give the simplest case, suppose I_1, I_2 are connected by causal process p_1, and I_2, I_3 are connected by causal process p_2. Suppose I_1 is defined in terms of the exchange of conserved quantity q_a, and I_3 by q_b Then I_1, I_3 are connected by a causal connection if I_2 involves the exchange of q_a and q_b, and p_1 possesses q_a and p_2 possesses q_b.

For the most general case of cause and effect we can write the cause as q(a) and the effect as q'(b), where a and b are objects and q and q' are conserved quantities possessed by those objects respectively. Then, to generalise our earlier formulation, we can say that:

Causal Connection: There is a causal connection (or thread) between a fact q(a) and a fact q'(b) if and only if there is a set of causal processes and interactions between q(a) and q'(b) such that:

(1) any change of object from a to b and any change of conserved quantity from q to q' occur at a causal interaction involving the following changes: $\Delta q(a)$, $\Delta q(b)$, $\Delta q'(a)$, and $\Delta q'(a)$; and

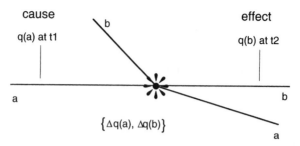

Figure 7.3. Collision between two balls, a, b. The momentum of the first, q(a), is the cause of the momentum of the second, q(b).

(2) for any exchange in (1) involving more than one conserved quantity, the changes in quantities are governed by a single law of nature.

The need for (2) is to rule out cases where independent interactions occur by accident at the same time and place. (No deep account of laws is required here – simple covariance will suffice. The possibility of worlds with accidental covariance is beside the point for this empirical analysis.)

Further, causation can occur at the manifest level. If c_m, e_m are manifest facts, and c_p, e_p are physical facts, then c_m causes e_m only if c_p causes e_p and c_m supervenes on c_p and e_m supervenes on e_p.

For example, to take a simple physical case, the earlier momentum of a billiard ball (q(a) at t_1) is responsible in the circumstances for the later momentum of the same ball (q(a) at t_2). Then the cause-fact and the effect-fact are linked by a single thread involving just the object a and the quantity q, momentum.

As a second example, shown in Figure 7.3, suppose the ball collides with another, so that the earlier momentum of the first ball (q(a) at t_1) is causally responsible for the later momentum of the second ball (q(b) at t_2). Then the cause and the effect are linked by the 'thread' involving first q(a), the first ball having a certain momentum; then the interaction $\Delta q(a)$, $\Delta q(b)$, the exchange of momentum in the collision; and then q(b), the second ball having a certain momentum. There is a change of object along this 'thread,' but no change in the conserved quantity.

As a third example, suppose an unstable atom is bombarded by a

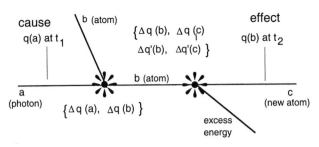

Figure 7.4. Incident photon a strikes atom b, raises its energy level, which then decays to atom c.

photon of absorption frequency, which leads to the subsequent decay of the atom (Figure 7.4). Take q to be energy and q' to be charge. Then the cause q(a), the incident photon with certain energy, is linked to the effect q'(c), the existence of the second atom, the product of the decay. It is linked by a thread involving one interaction where there is an exchange of energy, and a new object; and a second interaction where one object becomes another, with an exchange of charge and of energy, which leads to the effect-object having a certain charge, by virtue of which it is called the decay product. Further, there are laws governing how energy changes in such a decay.

Suppose now, to vary the example, that the atom happens not to decay. Then, by my account, the incident photon is not the cause of the fact that the atom did not decay, since the effect concerns charge, not energy, and there is no interaction where both energy and charge are exchanged.

We can now see how to deal with misconnection cases such as 'the tennis ball's momentum causes it to be green.' This is ruled out because there is no appropriate thread involving an object and a conserved quantity. There is the continued existence of the ball and its momentum, but that does not belong to the same thread as the existence of green fur. The ball's being green at t_1 is no cause of the ball's having the momentum it does at t_2, because, while there is a 'thread' linking the fact that it is green at t_1 with the fact that it is green at t_2, and a separate thread linking its earlier and later momentums according to our definition, there is no thread that fits that definition and that links the ball's being green at t_1 with its having a certain momentum at t_2.

173

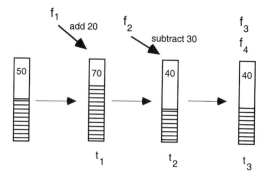

Figure 7.5. Disjunctive relata.

It also solves the case of misconnection where my tapping the desk, although causally irrelevant to the clock hand moving shortly after, is connected by sound waves or just a series of atoms colliding. The alleged effect, the hand moving, no doubt supervenes on a genuine physical fact involving a conserved quantity. But it is not connected to the sound waves, since whatever quantity is involved in the string of processes and interactions that constitute the sound waves, it is not the quantity that changes in the interaction underlying the 'hand moving.' If, perchance, the motion of air molecules did make some slight difference to a distant event, then that would qualify as a (minor partial) cause.

This also solves the sprayed plant case, if, for example, the cells affected by the poison simply die and the plant's continued life is the result of other independent cells.

However, this is a special case. To see how this account handles these sorts of counterexamples in a more general fashion, suppose that some object has quantity $q = 50$, and that a critical value of q for the object is 40 (below which it decays, or is dead, or something). Suppose an interaction occurs that raises the value of q by 20, then a second interaction occurs that lowers the value of q by 40, leaving it with a q-value of 30, below the critical value. Define the events (see Figure 7.5) as follows:

f_1 – q's value was raised by 20 at t_1
f_2 – q's value was lowered by 30 at t_2
f_3 – q's value was 40 at t_3
f_4 – q's value was less than 50 at t_3

Now, we might be happy to say that f_1 caused f_3, because without f_1 the object would have had a lower value of q (although this would be debatable). But we would not, by any stretch of the imagination, say that f_1 caused f_4. For a start, without f_1, f_4 would have been much more likely. But the problem is that f_1 *does* cause f_4, because it's the same object and the same quantity involved in both events, connected by a single causal thread. It seems to me that this captures the basic difficulty behind cases such as the sprayed plant and the fat child.

An example of this would be where, in a cold place, the heater is turned on for an hour, bringing the room to a bearable temperature. But an hour later the temperature is unbearable again, say 2°C. Then on my account the fact that the heater was turned on is the cause of the fact that the temperature is unbearable at the later time.

However, we have seen that disjunctive facts are ruled out as candidates for genuine physical causal facts. We have also seen that facts like f_4 are disjunctive. So there cannot be a causal connection at the physical level between f_1 and f_4. Given this, there is no connection at the manifest level, either.

In general, that c causes e does not entail that c causes d where e entails d. This seductive notion needs to be resisted.

VII.5 SUMMARY

It would seem, on the account offered in Chapter 5, that items not related causally can be connected by sets of causal processes and interactions – the problem of 'misconnections.' We considered how chance-raising might be added to the theory so as to solve this problem and the main problem facing probabilistic theories, that is, chance-lowering causes (Chapter 2). We presented a solution that showed how the thesis that causes are chance-raising can be saved, but we found other difficulties with this solution. Finally, we showed how appropriate restrictions on the kind of facts that can be physical causes and effects, and on the way that sets of causal processes and interactions can connect them, solves the problems of misconnections. Thus we have arrived at a satisfactory answer to our second question, *what is the connection between causes and effects?*

8

The Direction of Causation and Backwards-in-Time Causation

I have claimed that it is profitable to distinguish three key questions about causation. The first question, *what are causal processes and interactions?*, was addressed in Chapter 5. The second question, *what is the connection between causes and effects?*, was addressed in Chapter 7. The third question, *what makes a cause different from an effect?*, has not been addressed, and it is to that task that we turn in the present chapter.

The question of the directionality and asymmetry of causation has long been an important ingredient in philosophical discussions of causation, although perhaps Reichenbach (1991) was the first to truly appreciate the significance of the question. It is certainly an aspect of folk intuitions. Many philosophers have held that an adequate analysis of causality should provide an explanation or account of causal asymmetry. On the other hand, it is precisely this aspect of causation that is the most common target for elimination from empirical accounts. This chapter looks at the ability of 'process' theories of causality to provide such an account, granting certain constraints.

The major constraint, or premise, in the following analysis is the assumption that quantum mechanics reveals the presence of backwards-in-time causation. Thus, in the first instance, the argument presented here will not convince those who hold that there can be no backwards causation. However, I derive an empirical prediction, which, if confirmed, would give those sceptical about backwards causation reason to think again.

The chapter begins with an analysis of some of the formal properties of causation. This is followed by a nontechnical presentation of the famous difficulties associated with Bell phenomena in quantum mechanics, and the backwards-in-time solution to those difficulties. This model is then used as a premise with which to assess the various philosophical theories of causal asymmetry. In section 8.4 the temporal theory is discussed, and in section 8.5 the subjective theory is discussed. Both of these are rejected as incompatible with backwards causation. We then turn to the fork asymmetry theory. Two common versions are distinguished, and each is shown to face difficulties. Another account, the kaon theory, faces similar difficulties. It is then argued in section 8.9 that a third version of the fork theory avoids these difficulties. Finally, an empirical prediction is derived from the conjunction of the backwards causation Bell model and the third version of the fork theory. I conclude that this third version of the fork theory is the correct account of causal asymmetry.[1]

VIII.1 ASYMMETRY AND DIRECTION

The causal relation is generally held to have a direction, since it is asymmetric. What does this mean, and is it correct? A two-placed relation $R(a,b)$ is *symmetric* iff $R(a,b) \supset R(b,a)$; *asymmetric* iff $R(a,b) \supset -R(b,a)$; and *nonsymmetric* iff it is neither symmetric nor asymmetric.

The singular causal relation, which relates a particular cause to its effect, is almost universally thought to be asymmetric. It seems not to be symmetric, since 'that my hitting the glass with the hammer caused it to break' does not entail 'that the glass breaking caused my hitting it with the hammer.' If the singular causal relation is asymmetric, then it clearly has an objective direction. For the arrow of causation points from the cause to the effect, and that direction is fixed by virtue of the relation obtaining, since if a causes b, then the causation occurs 'from a to b,' and not 'from b to a.'

But could the singular causal relation be nonsymmetric, like the relation 'reminds me of'? Is it possible that a particular effect is the cause of its cause? One argument that it is nonsymmetric is the two planks argument. Two planks of wood are leaning on each other, like an upside down V. Is not the fact that one plank is standing in that position a (partial) cause of the other plank's standing in its position; and is not the converse also true? Various responses are available here. One is that such facts are really time-indexed,[2] so that it's really the posi-

---

1. In doing so, we will need to skip quickly across two areas of debate: first, the interpretation of quantum mechanics; and second, the fork theory of the direction of causation. For a more detailed analysis, see Dowe (unpublished ms).
2. This view is expressed in Menzies (1989b).

tion of plank A at an earlier time that is the cause of plank B's position now, which in turn is the cause of plank A's position at a later time, not at the earlier time. Nevertheless, we can see that the point is debatable. A second argument that the singular causal relation is nonsymmetric appeals to an even more contentious issue as to whether singular causation is reflexive, irreflexive, or nonreflexive. This argument claims that singular causation is nonreflexive, since, for example, God is His own cause, yet most causes are not self-causing. If this is true, it follows that causation cannot be asymmetric, because at least some effects cause their causes, namely those where the effect and the cause are identical.

If the notion that singular causation is nonsymmetric seems crazy, it may be worth mentioning that an even stronger case can be made for the thesis that general causation, the relation between types of events, is nonsymmetric. Davis has argued that general causal laws are nonsymmetric (1988: 146). He gives the example of a husband and wife who regularly give each other colds. Each person's having a cold can be said to be the cause (type) of the other's having a cold. Thus if F is 'the wife's having a cold,' and G is 'the husband's having a cold,' and C is the general causal relation, then $C(F, G)$ and $C(G, F)$ both hold; therefore $C(F, G)$ is in general not asymmetric. A similar example is due to von Wright: Rain causes flooding, and flooding causes rain. It's easy to see how to generate this type of counterexample using any of those situations that we call vicious circles, or feedback systems – for example, price increases cause wage increases, and wage increases cause price increases.

Eells disagrees (1991: 240). Appealing to a principle (due to Nancy Cartwright) that in general $C(F, G)$ holds only if we hold all other causally relevant factors fixed, Eells says that a factor that must be held fixed is the time at which the event obtains. If F_1 is 'the wife's having an earlier cold' and F_2 is 'the wife's having a later cold,' and similarly for G_1 and G_2, then $C(F_1, G_2)$ and $C(G_1, F_2)$ hold, but $C(F_2, G_1)$ and $C(G_2, F_1)$ don't; so causal laws do turn out to be asymmetric.

I think this is a mistake that subtly confuses the type and the token. For conjoining $C(F, G)$ and $C(G, F)$ is not a claim for backwards-in-time causation. That would be so if we conjoined $c(Fa, Ga)$ and $c(Ga, Fa)$, where c denotes the singular causal relation. When we conjoin $C(F, G)$ and $C(G, F)$, we are not claiming that the husband's cold caused the wife's later cold (as Eells implies) or that it caused the wife's earlier cold (as Eells prefers) but that *in general* husbands' colds cause wives' colds. It is a claim about types that doesn't need cases of backwards causation to make it true. What Eells does show is that general causes don't violate temporal priority. So the vicious circle argument still stands: causal laws are not asymmetric.

But, to return to the main line of argument, even if singular causation is nonsymmetric, it still has a direction. My argument for this key point is as follows. Suppose the singular causal relation between a and b, c(a, b), is nonsymmetric. Then for any given c, it is contingent whether c(b, a) or ~c(b, a). That is, it is established that the causal relation obtains in one direction, but it is not established for the other direction. Suppose, for example, that a is earlier than b. Then we know that the earlier event causes the later event, but we do not know whether the later event causes the earlier. We do not know whether the causal relation is bidirectional or unidirectional. But, of course, to be unidirectional is to have one direction, and to be bidirectional is to have two directions. Therefore, either way, the causal relation has a direction.

The question of the asymmetry of the causal relation, which is a logical property of a relation, is not to be confused with the question of the *temporal* asymmetry of causation. The temporal asymmetry of causation refers to the fact that causes always or nearly always precede their effects in time, sometimes called 'priority.' To see the difference, it helps to recognise that the logical asymmetry of causation does not guarantee the temporal asymmetry of causation, and that the temporal asymmetry does not guarantee the logical asymmetry (since there would be a temporal asymmetry if most but not all effects occurred after their causes, and where an occasional cause is caused by its effect).

One philosophical theory of the direction of causation holds that the difference between cause and effect is just that the former occurs before the latter. Call this the *temporal* theory of the direction of causation. If this is true, then we have a nice explanation as to why causes always precede their effects.

But this is not the only theory of the direction of causation. Another philosophical approach is to find the direction of causation in other physical facts, such as the contingent fork asymmetry: open forks are always open to the future. Call this the *physical* theory of the direction of causation. Other candidates for a physical basis of causation's direction include the kaon arrow. A third kind of approach we may call the *subjective* theory: the direction of causation is not an objective feature of the world, but reflects a perspective from which one views causation.

However, it is even more difficult to say in what sense a causal process has a direction. The ordinary concept of a process does involve a direction. If we speak of the process of an apple decaying, or an atom decaying, we are thinking of the process 'moving' in a direction, of its having a starting point and a culmination point. But, if a process is just a definite wormlike region in spacetime, as I argued in Chapter 5, then we must ask, in what sense may a definite wormlike region in spacetime have a direction? We cannot say that regions in spacetime 'go' or 'move' in any direction; that doesn't make sense.[3] The problem is particularly acute when one considers the temporal symmetry of the laws of physics, a consideration that has led some scientists and philosophers to be eliminativists about the direction of causation. The primary task of this chapter is to address this problem about the direction of causal processes.

VIII.2 BELL PHENOMENA AND BACKWARDS CAUSATION

I want to begin first by outlining the central puzzle in the interpretation of quantum mechanics, which lies at the heart of the so-called Bell phenomena, and second by presenting what I take to be the most promising solution to the puzzle, namely the backwards causation model. This solution is not widely accepted, mainly, it seems to me, because of some supposed philosophical difficulties, which I cannot address here; so I'm not expecting that by itself the argument of this chapter will convince anyone to accept the solution. In fact, in this chapter I simply take it as a premise that this model is the true explanation, in order to show what follows from it. But just to make it clear where I'm coming from: I do think that this is a very promising suggestion, I think that the supposed philosophical difficulties are very much overrated,[4] and I hope that the argument of this chapter will lead to the model's being taken more seriously.

Instead of discussing the physics of this remarkable phenomenon, I propose to illustrate the central puzzle by constructing an analogy concerning identical twins. This analogy is an imperfect analogy; I've deliberately simplified things in order to focus on the central puzzle. I hope to show that one does not need to know anything about physics in order to understand the really puzzling feature of the Bell phenomenon.

---

3. Compare Jack Smart's point about the flow of time (1963: 136).
4. See the excellent discussion in Price (1994).

Suppose there are identical twins from Sydney, Steve and Mark. One of these twins, Steve, moves to Hobart, where he falls foul of a mysterious disease and dies. It is reported that this disease is triggered by excessively cold weather, and a person with the condition will almost certainly die if he is subjected to, say, two or three days of continuous cold. This exotic disease has become known as "freezerphobia." The other twin, Mark, stays in Sydney, where he is hit by a bus and dies (a detail which has no point of analogy in quantum mechanics).

Let's call putting a person in Hobart "the cold test" or measurement. Now, we've heard rumours from scientists that when a pair of twins go to Hobart, often they both die. In fact, it is reported that there is a one-to-one correlation, namely, that if two identical twins are both cold tested, then one dies if and only if the other dies. On this evidence it's reasonable to think that this disease is genetic, and that that's why we get this correlation between identical twins – simply because identical twins share the same genetic makeup. Therefore (we might reason, given what we know about Steve) we can infer that Mark also had freezerphobia, even though we can't actually test that, since he's dead, having been hit by a bus.

So we have formed a kind of theory, a mechanistic theory. We think the correlation is due to an underlying mechanism – genetic, probably – that can be traced back to a condition of the unsplit egg from which the twins were produced. That is, we suppose that the correlation has what philosophers call a common cause explanation (see Salmon 1984). It's a reasonable hypothesis, and the reasoning seems to be common sense.

Let's move now from this particular case of our friends Steve and Mark, and look instead at what the scientists are doing about this. In fact we find, already established at St. Vincent's Hospital in Sydney, the so-called Freezerphobia Register of Scientific Terminations or FROST for short. (The significance of the term 'terminations' will become clear later.) We find that the scientists have set up an experiment, and also that many of the scientists already have formed a theory similar to our own common sense idea of an underlying genetic mechanism, but, being good inductivists, they have waited to see how the experiment turns out before forming any firm beliefs about the matter.

Let's look at the experiment under four headings: the set-up, the method, the results, and the interpretation. First, the experimental set-up involves attracting pairs of identical twins from the general population to volunteer as participants. This is done by offering participants free trips to either Hobart or Dunedin, in New Zealand. Two planes have been chartered, and observation cells have been set up in Hobart, Dunedin and Sydney.

Second, the method has four steps. (1) A pair of identical twins are taken to Sydney Airport. One twin is put on the Hobart plane, and one on the Dunedin plane. When measures have been taken to ensure that there can be no further communication between the twins, both planes take off. (2) Once in the air, the Dunedin pilot tosses a coin: if it's heads, she heads for Dunedin; if it's tails, she does a big loop and lands back at Sydney Airport. (3) On arrival at whatever destination, the subjects are locked up in the observation cells and are carefully monitored. (4) The scientists simply record whether or not the participants die. These four steps are followed for large numbers of identical twins.

The results of the experiment are divided into two groups, because there are two possible experiments done on a pair: they could be taken to Hobart and Dunedin, or to Hobart and Sydney. Remember that it a matter of chance which of these two experiments is performed on a particular pair of twins. For the Hobart-Dunedin pairs, it is found that 62% of those who get to Hobart die, and 62% of those who get to Dunedin die. In each case the other 38% survive with no ill effects, apart from occasional nonfatal cases of frostbite, flu and chilblains. Further, it is noted that there is a one-to-one correlation with respect to the ones who died: the 62% who died in Hobart were the siblings of the 62% who died in Dunedin (the result I referred to earlier). We should note here that the results so far were regarded by most of the scientists (but not all) as confirming the genetic mechanism theory, because that theory predicts this one-to-one correlation. So far, so good.

But for the Hobart-Sydney pairs, the results stunned everyone. They found that 45% of those who end up in Hobart die, while none of those who end up in Sydney die. What's puzzling about that? What's puzzling is that 45% is not the expected result. One would have expected that figure to be about 62%, because that's what the genetic mechanism theory predicts. It should not make any difference to those heading for Hobart whether their twins go to Dunedin or Sydney. If it's already set by your genes that you will die if cold tested, then it is irrelevant whether or not your twin is cold tested. These results are indeed puzzling, and it is this feature that is parallel to the central puzzle in Bell phenomena.

So let's turn to the scientific interpretation of these freezerphobia results. First, the results are a major setback for the genetic mechanism theory, at least in its present form. In fact, the scientists all agree that the genetic mechanism theory in its present form is refuted by these data, and being good Popperians, they abandon that theory. But with what can it be replaced? Well, when the dust settled there were numerous differing views amongst the scientists, but there were three major positions that attracted support.

One group, called the Copenhagen school, focussed on the mathematical description of the results of the various measurements. They held that we should simply figure out the mathematical formula, called the state function, that describes all the results of all the measurements, both actual and possible. But we should not ask how to explain it. There is no explanation. The state function should be regarded as a mathematical tool for making predictions about the results of measurements, and not as a pointer to underlying mechanisms. "The state function is complete," was their slogan. Actually, they had not been pleased with the way most scientists had interpreted the first part of the experiment, and were feeling pretty smug about the way the second part had turned out. In fact, they now claim that the results prove that the state function is complete.

A second group of scientists were unable to accept the position of the Copenhagen school, feeling that there was a genuine need for science to *understand* what is happening. So they came up with the telepathy theory. According to this theory, there's a nonmechanistic communication going on between the twins. A twin *feels* the death of the other twin even though the twin is miles away, and in sympathy or harmony curls up and dies with him. It's not by conscious decision, but by an unconscious harmonic necessity. This nonmechanistic communication doesn't always work, but it works enough to have an effect. In fact, this group, who became known as the 'telepathists,' regarded the results of the experiment as clear evidence of such telepathy.

A third group of scientists, who like the telepathists had a keen desire to find understanding, but who disliked the New Age connotations of the telepathy theory, suggested an explanation called the backwards causality model. According to this theory, the condition freezerphobia in the undivided cell in the mother's womb is in part caused by future events, for example, the future event that a twin is subjected to the cold test twenty years later. That explains why the rates of death in Hobart seems to depend on whether the twin went to Dunedin or not. One's twin being cold tested in Dunedin, an event

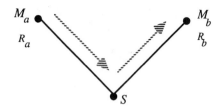

Figure 8.1. The backwards causation model.

caused by the coin toss, was the cause of the earlier genetic condition, which in turn was the cause of the Hobart twin's death (see Figure 8.1). Choice of measurement on one twin causally influences the earlier state, and the earlier state causally influences the result of the measurement on the other.

The opponents of this theory claimed that it leads to paradoxes. For example, they said, if you could cause past events then you could shoot your grandmother before your mother's birth, and then where would you be? Both not alive and alive at the same time, which, they correctly observed, is quite impossible. Others, mainly from the Copenhagen school, said that backwards-in-time causality is simply an abuse of language. The very meaning of the words 'cause' and 'effect' involve the ideas of 'earlier' and 'later' respectively. So there was no general agreement, and a period of Kuhnian crisis set in, which remains to this day.

The situation in quantum mechanics is analogous to this parable in a number of important ways. In particular, the situation in Bell set-ups seems to be such that the predictions of so-called local realist theories[5] come into conflict with the predictions of the quantum mechanical state function. But the local realist theories can be derived from very plausible assumptions. This difficulty, first articulated by the Scottish physicist John Bell in 1964, is perhaps one the most startling conceptual problems in the history of science. As is well known, experiment has clearly supported quantum mechanics.

Consider a system involving two particles, A and B, separated from a singlet state S. Suppose there are several possible measurements one could subsequently perform on these particles, and that a different measurement is performed on each of the pair. We may treat the actual measurements as causal interactions between the experimenter

5. See the presentation in d'Espagnat (1979).

184

and the particle. Call these interactions M_a and M_b, and the results of these measurements R_a and R_b respectively. Leaving aside numerous niceties, relevant experimental results entail that one of the following two propositions is false:

1. There is no causal connection between the act of measurement on A and the result of the measurement on B, that is, between causal interactions M_a and R_b.
2. The measurement reveals an underlying reality about the particle, which exists independently of the measurement.

Proposition 2 is important because it entails the truth of certain counterfactuals of the form:

2′. If the alternative measurement had been performed on A, the result would have been _____,

where 2′ is deduced from known correlations between A and B and the result of the measurement on B, together with assumptions 1 and 2. Bell's reasoning then deduces from 1 and 2′ predictions that contradict both the quantum mechanical prediction and the experimental results. So it is thought that either 1 or 2 must be abandoned; but the problem is that neither avenue is particularly appealing, for a variety of well-known reasons.

There are numerous suggestions as to how to solve the puzzle. The backwards causality model, although not widely supported, is in my view one of the most sensible solutions.[6] It suggests a physically well-defined mechanism by which one measurement can influence another. Essentially, the idea is that the act of measurement on particle A brings about causal influence that propagates backwards-in-time to the source of the two particles, whereupon it is partially causally responsible for some hidden characteristics of the state of that pair of particles. Those characteristics are in turn causally responsible for the outcome of the measurement on B.[7] Thus the difficulty is resolved by the hypothesis of backwards causation.

---

6. The telepathy theory has its parallel in suggestions that there is some faster-than-light influence between two well-separated measurements. This is problematic because it contradicts the Special Theory of Relativity.
7. See the account in Sutherland (1983; 1985).

One promising version is the 'transactional' interpretation of quantum mechanics, due to Cramer (1986; 1988a). On this account, quantum events are understood as causal interactions between retarded waves travelling forward in time and advanced waves travelling backwards-in-time (which usually are discarded as unphysical). Cramer describes the transaction, or 'handshake,' as follows:

the emitter produces a retarded offer wave (OW), which travels to the absorber, causing the absorber to produce an advanced confirmation wave (CW), which travels back down the track of the OW to the emitter ... The exchange then cyclically repeats until the net exchange of energy and other conserved quantities satisfies the quantum boundary conditions of the system, at which point the transaction is complete. (1986: 662)

This transactional interpretation provides an objective interpretation of the quantum mechanical wavefunction, and Cramer argues that it resolves various quantum paradoxes (1986: sec. 4). The price paid is that it postulates causal influence from the future to the past. Given that the benefit is such a remarkable achievement, that price is a price that we would like to be in a position to pay.

Bell phenomena fit easily into the 'process' way of thinking. The world line of the particle as it moves from the source to the measurement is a process;[8] the world line of the other particle is another process; while the separation at the source is an interaction. The measurements should also be taken as interactions. It is clear that if the backwards causation model is to work, there *must* be a sense to the idea of the direction of a causal process. For the suggestion is that the causal influence goes, via the causal process, in a direction contrary to the usual direction of causal influence. There must be an objective difference between the directions of the two processes.

VIII.3 THE PREMISE

For the sake of the argument of this chapter I shall take it that this hypothesis of backwards causation is true, and that in fact this is precisely what happens in Bell phenomena. The purpose is to show what follows from such an assumption for philosophical theories of the

---

8. When we say that causal influence moves from the measurement back to the source, we may mean either that of the process has two directions, or that there are two coincident processes with opposite directions, occupying the same series of spacetime points. Which conception is preferable, I cannot say.

direction of causal processes. Thus the premise of my argument is the backwards causation model, and the conclusion is a philosophical theory about causal asymmetry.

In section 8.2 reference was made to a number of philosophical theories about the direction of the causal relation. In what follows I will adapt these theories to the notion of a causal process, to see whether any of them can make sense of the direction of a causal process in a way that is compatible with the backwards causation model of quantum mechanics. That is, I will not be attempting to prove anything about the direction of causal processes per se. I am just starting with a premise, viz., the truth of the backwards causation model, and I seek to determine what follows from that premise concerning the direction of causal processes. I argue that some very unusual conclusions follow, about an empirical prediction capable of confirming the backwards causality model.

My argument, then, clearly is open to being refuted: if the predictions prove to be false, then we may be forced to take all this as an argument against the backwards causation model. Philosophers who in any case find the notion of backwards-in-time causation repugnant on other grounds may prefer to take my argument as a reductio of the model. My only response to this is to point out that if the predictions suggested here were confirmed, that would provide grounds for sceptics to think twice. In the absence of such evidence, I prefer simply to direct my argument towards those who are attracted to the idea of backwards causality, and to examine the implications of that kind of commitment.

VIII.4 THE TEMPORAL THEORY

As we have seen, there are three kinds of theories of causal direction: the temporal, the subjective and the physical. The temporal theory is widely held amongst philosophers, but is easily dismissed in the present context. The temporal theory of the direction of causation asserts that the direction of causation is defined by the direction of time. David Hume held this view, as we saw in Chapter 2. According to Hume, two objects A and B are related as cause and effect respectively only if A is precedent to B (*Treatise*: 169). Thus, by definition, the direction of causation is the direction of time, as a matter of conceptual necessity. An effect cannot be prior to or simultaneous with its cause. The Humean position is a common one in philosophy. We have already seen that Suppes and Fair also define causal asymmetry in terms of temporal priority.

However, it is clear that the temporal theory is incompatible with the backwards-in-time model. By definition, it is impossible for a later measurement to have causal influence on the earlier state of the particle pair. Therefore, given that we are seeking to understand the direction of causation in a fashion adequate to the explication of the backwards causation model, we must rule out the temporal theory.

Compatibility with backwards causation is a requirement that many philosophers now demand, chiefly on account of suggestions that backwards causation may one day find a place in physical theory. Discussion of tachyons, Feynman electrons, and (as we have seen) the backwards causation theory of Bell phenomena all point in this direction, despite a long tradition of philosophical argument to the contrary. But the feeling is that it will not do for the philosopher to rule out a priori what the scientist is currently contemplating as a serious hypothesis. Other philosophers take this as an indictment of those scientists.

So the theories we shall take seriously are the subjective and the physical. The subjective theory attempts to explain causal asymmetry as something we add to the world. The physical theory attempts to reduce causal asymmetry to something physical in the world. We shall deal with these in turn.

VIII.5 THE SUBJECTIVE THEORY

According to the subjective account, causal asymmetry is not a feature of the world of objects, events and processes but rather of the way we see those objects, events and processes. Subjectivism comes in two forms. A stronger form holds that causation itself is not an objective matter. An example of this stronger form is the manipulability account of causation, which holds that causation is to be analysed in terms of agency, the human capacity to act to make a difference in the world.[9] A weaker form holds that while causation itself is objective, the direction of causation is subjective. An example of this weaker form is the eliminativism of Wheeler and Feynman, who, on account of the symmetry of the laws of nature, regarded as physically equivalent the statements,

---

9. See Gasking (1955) and von Wright (1971). For a recent defence see Price (1991) and Menzies and Price (1993).

"The glass broke because it was hit with a hammer," and "The hammer hit the glass because it was going to break it" (cited in Horwitz et al. 1988: 1161).[10] The arguments presented here will apply to either form.

Subjectivism has recently been defended by Huw Price, under the label of 'perspectivism.' According to Price,

> in a certain sense causal asymmetry is not in the world, but is rather a product of our own asymmetric perspective on the world. We ourselves are strikingly asymmetric in time. We remember the past and act for the future, to mention two of the more obvious aspects of this asymmetry. It does not seem unreasonable to expect that the effects of this asymmetry will come to be deeply entrenched in the ways in which we describe our surroundings; nor that this entrenchment should not wear its origins on its sleeve, so that it would be easy to disentangle its contribution from that part of our descriptions we might then think of as perspective-independent. (1992: 513–514)

It is not obvious to us that causal asymmetry is subjective, and intuitively we might feel that it is an objective matter. But if we compare the case of colour, Price argues, we will recognise that sometimes a perspectival aspect of the world will not immediately be recognised as such. Before the seventeenth century, the common view of colour was that it existed wholly in the objects, and it took advances in science and philosophy to recognise that there is a subjective aspect to colour.

On the subjective account, backwards-in-time causation may well be possible. If we could view the world differently, so that we saw processes as going backwards-in-time, or causes occurring after their effects, then that would be backwards-in-time causation. However, the kind of backwards causation required in the backwards causation model of Bell phenomena is not possible on the subjectivist account. For that model to work, the distinction between the direction of process M_a-S and the direction of process S-R_b must be objective, because subjective aspects cannot appear in basic physical theories of the objective world. If causation's direction is but a function of the way we experience the world, then we cannot solve paradoxes in the quantum realm by positing it there.[11]

10. Earman (1976) also seems to take this view.
11. There are attempts to use the notion of subjectivity to solve quantum dilemmas; but such attempts do not need backwards causation. There's no need to introduce two conceptual difficulties just because one (or other) seems necessary. For a discussion of one attempt to mix subjectivity and backwards causation, see Dowe (1993a).

Consider, for example, the disanalogy to colour. One plausible account of colour says that part of the notion of colour is subjective, but that part never appears in physical theories, for example, in surface physics. It is precisely because colour does not appear in the physical theory of the objects that we think of it as subjective. Indeed, it is *only* because colour does not appear in the physical theory that we think of it as subjective. Had a full physical reduction of colour been available, the issue of colour being a secondary quality would not have arisen, because objects would have been red independently of anyone's perceiving them and colour would have been taken as objective. But colour is subjective, and a world without observers of any kind would not be a coloured world. On the other hand, causal direction *does* appear in physical theory, so *it* cannot be subjective. If the backwards model is correct, then even if there never were observers of any kind, there would still be backwards-in-time causation whenever Bell phenomena occur.

Price has offered a detailed defence against this kind of objection (1994; 1996a). Price argues that even though causal asymmetry is perspectival, it is nonetheless 'objectified,' in the sense that we are not normally free to see things any other way. The reason is that we deliberators are orientated one way in time, in the sense that we take the input for deliberation from the past, not the future, as if the past is fixed and the future open. Nevertheless, there may be circumstances in which we could deliberate in the reverse direction. These circumstances would have to be such that we could not possibly know about the earlier effect of our later action until that action is completed; otherwise, on seeing the earlier effect we could decide not to carry out the cause (so-called bilking).

However, this is exactly the situation we find in the Bell case. For the only way we can know about the relevant hidden facts about the earlier state S is via a measurement such as M_a or some mutually exclusive alternative to M_a. Thus it is physically impossible to know the hidden part of S in time to decide to change the cause, M_a.

It follows, Price claims, that there are two possible objective microstructures that have different implications for how we may view the direction of causation. Normal, 'classical' structures do not rule out bilking and thus do not permit a backwards interpretation. But Bell correlations, between, say, the hidden parts of S and M_a, do allow a backwards interpretation. Thus the 'objective core,' as Price calls it, of the backwards-in-time interpretation is just this correlation between

M_a and S, that is, between an earlier hidden state and a later choice of measurement.

However, it seems to me that this fails to answer the objection. We should ask the question, are we deliberators *forced* to interpret this correlation as backwards causation? What I mean by this is: are we forced in the same way that under normal conditions we are not free to see things other than in the usual direction, so that, under these special conditions, it is the case that we are not free to see things other than in a backwards direction? If the answer is no, as Price appears to hold, then it's hard to see how the process M_a-S has the requisite objective direction. That objective facts make it *possible* to interpret the process as going backwards-in-time, doesn't mean that interpretation gains the status of 'objective.' Clearly, if both a backwards and a forwards interpretation are possible, then the objectivity of the backwards interpretation is not secured. It's not even 'objectified' in Price's sense.

In fact, such a correlation does leave open a choice. We could see it as (1) backwards causation, (2) normal causation, where the hidden state directly causes the choice of measurement, or (3) so-called cryptodeterminism,[12] where S and M_a have a common cause; for example, something in the experimenter's past that determines his choice and that also brings about the hidden state S. In other words, the objective core by itself does not have any implication for the direction of causation; and the bilking loophole simply makes backwards causation possible. Thus the objective facts do not serve to necessitate or 'objectify' the backwards perspective.

On the other hand, suppose the answer is yes, that is, we are forced to interpret the correlation as backwards. Even then, the defence would not work. For we must now ask the question, is our perspective here objectified in full by features of the physical system? If the answer to this second question is no, then the objectivity uncovered is not objectivity enough. Consider again the analogy to colour. Suppose physicalism is correct, and our experience of seeing red the way we do is a product of our hard-wired physiology. Then we humans are forced to see a certain kind of object in certain circumstances as red, which means that our experience of seeing red is 'objectified' in Price's sense. But being red remains a subjective matter, since being red is not a pure

---

12. To borrow, and change, Belinfante's terms, see Belinfante (1973).

quality of the physical objects. So this kind of 'objectivity' is not the kind of objectivity required for a physical theory. Perspectivism 'objectified' is not objectivity enough.

But if the answer to this second question is yes, that is, in Bell cases our perspective is objectified in full by features of the physical system, then it would seem that causal direction could not be perspectival after all, at least not in this case. For something to be perspectival it must be added by us, even if it is 'objectified' in the sense that we have no choice but to adopt that perspective. But in this case, causal direction is not added by us but is completely given by features of the physical system.

So it appears that there is no way to adequately objectify perspectivism. If this reasoning is correct, then the truth of the backwards-in-time model does rule out the subjective account of causal direction, as our initial argument claimed.

VIII.6 THE PHYSICAL THEORY: FORK ASYMMETRY

The idea of fork asymmetry was first articulated by Hans Reichenbach in his book *The Direction of Time* (1991). Reichenbach begins with the notion of a causal fork. A causal fork may be defined as a set of three events, two of which are the joint effects of the third, the common cause. For example, the presence of a virus in a heated room is the common cause of the joint effects where two brothers both get the flu. In terms of processes, we may define a causal fork as the case where two causal processes arise out of one causal interaction.

Joint effects of a common cause tend to be correlated in one way or another. They bear the marks of their common causal ancestry. The effects of the virus on one brother tend to be similar to the effects of the virus on another. If one wishes to analyse such correlations, one must turn to statistical relations, said Reichenbach. This brings us to the question of conjunctive forks.

Simplifying for the sake of exposition,[13] a conjunctive fork may be defined as a set of three events A, B and C such that

$$P(A.B) > P(A)P(B) \tag{1}$$

$$P(A.B|C) = P(A|C)P(B|C) \tag{2}$$

---

13. The full characterisation can be found in Salmon (1984: 159–160).

For the sake of the present discussion we may follow Reichenbach in taking these probabilities as relative frequencies. Equation (1) expresses a statistical correlation between two events, namely, that they are not independent. If the joint probability were equal to the product of the separate probabilities, then they would be independent. Equation (2) expresses what Reichenbach calls a 'screening-off' relation, where a third event C is found that when accounted for, renders A and B independent. Together, such a set of events form a conjunctive fork. The significance of the conjunctive fork, in Reichenbach's view, is that we can characterise causal forks statistically, by the conjunctive fork.

For example, suppose the probability of a person taken randomly from the general population having the flu at any given time is 1/10, and that the joint probability of two members of one family having the flu is 1/60. Then we have an example of the correlation depicted in (1) (since 1/60 > 1/100). Thus (1) gives us a precise way to express the idea of a correlation between separated events. However, if the sole reason that two brothers have both caught the flu is that they were both in a heated room where a virus was present, then the fact C, that they were both together in a heated room where a virus was present, will screen off the correlation, according to (2).

We can take it as read that A and B are roughly simultaneous, but we haven't said whether the common cause C is in the past or the future of the events it screens off. Can we in fact find a conjunctive fork where C occurs after A and B? The answer is yes.

Suppose instead of the flu, the brothers catch a rare and dangerous viral disease, RDV. Call these unfortunate events A′ and B′. Then, again,

$$P(A'.B') > P(A') P(B')$$

Suppose a short time later they both end up the RDV unit in the local hospital, event E. Since this is the only kind of disease that is handled by this unit, the probability that a person has contracted RDV given that they are later in the RDV ward is 1. So

$$P(A'.B'|E) = P(A'|E) P(B'|E) = 1$$

This is equation (2); that is, we have another conjunctive fork, where E screens off the earlier correlation between A′ and B′. So we can have a conjunctive fork where the screening-off event occurs after the correlated events.

But notice that this is the second screening-off event for that correlation, since, just as with the flu, the correlation would also be screened off by the earlier event C′ – that the brothers were in a heated room where the virus was present. So we have what Reichenbach called a 'closed fork,' that is, a fork with screening-off events in the past and in the future. A fork not screened off in both directions Reichenbach calls an open fork.

Conceptually, open forks can be open either to the future or the past. However, Reichenbach pressed the following claim:

Fork Asymmetry Thesis. There are no conjunctive forks open to the past.

This is a claim about the way the world actually is, contingently. Reichenbach claimed that we find conjunctive forks that are closed, we find conjunctive forks that are open to the future, but we never find conjunctive forks that are open to the past. Correlations in fact are never screened off by a future event, except where they are also screened off by a past event. Will we find a conjunctive fork where C occurs after A and B *and* there is no screening off event before A and B? Reichenbach's answer is no.

We now turn, as Reichenbach did, to an account of the direction of causation. Roughly, the idea of a fork asymmetry account of the direction of causation is that the direction of causal processes is given by the direction of open conjunctive forks. Then we may call the screening-off event of an open fork a 'cause' and the events screened off 'effects.' This, together with the asymmetry thesis that all (or most) actual open forks are open to the future, explains why causes always (or mostly) precede their effects.

But this rough idea is ambiguous. In fact, there are at least three possible ways this could be taken. The first is how some contemporary philosophers explicate it; the second is Reichenbach's version; and the third is the version that I wish to articulate and defend as the best available account of causal asymmetry. It is important to distinguish these various versions, because, as I shall argue, they are open to different kinds of objections.

VIII.7 FORK THEORY VERSION ONE

Version one goes as follows: The direction of a causal process is given by the direction of an open conjunctive fork part constituted by that process. This is essentially the version defended by Papineau, except that Papineau discusses events rather than processes:

(SO) Take any event C. Then among the events which are correlated with C will be some that are correlated with each other in such a way that their correlation is screened off by C – these are C's effects; and among the events which are correlated with C will also be some that are not correlated with each other – these will be C's causes. (Papineau 1993: 239–240)[14]

This account of causal asymmetry has recently been subject to close scrutiny by Huw Price. According to Price, the fork asymmetry is "not a sufficiently basic and widespread feature of the world to constitute the difference between cause and effect" (Price 1992: 502).[15] This argument is claimed to work equally well for other related physical asymmetries, such as Popper's asymmetry of radiation – wave phenomena such as ripples on a lake radiate out from a central source but never converge into a central sink; and Lewis's asymmetry of overdetermination – in general, an event has numerous future determinates (events minimally sufficient for that event, given the laws of nature), but rarely has more than one past determinate.

The first step in Price's argument is to claim that the relevant asymmetry (fork, radiation, or overdetermination) is thermodynamic in origin; that is, the asymmetry arises because of the de facto thermodynamic disequilibrium that happens to obtain at our stage of the universe. In other words, the various physical asymmetries in question all reduce to the thermodynamic asymmetry, that the amount of entropy in our universe is increasing on account of the fact that we are moving out of a low-entropy, or highly ordered, initial state.

The second step in the argument is to point out that the thermodynamic asymmetry is a macroscopic phenomenon, which is not present, or 'visible,' at the microscopic level. The laws of nature that govern the motion of molecules are completely symmetric with respect to time, and it's only at the macroscopic level that we see irreversible processes – processes that involve a transition from low-entropy initial conditions to higher-entropy later conditions. This is because entropy is a property of a macrostate and is defined in terms of the number of possible microstates that correspond to that macrostate, relative to the total number of possible microstates.

14. See also Hausman (1984).
15. See also Price (1993: sec. 4).

Therefore, if causal asymmetry is grounded in thermodynamic asymmetry, then there can be causal asymmetry only at the macro level. In other words, these physical theories of causal asymmetry entail that there is no asymmetric causation at the micro level. This is unacceptable for two reasons, Price argues. First, it conflicts with reductionist intuitions to the effect that macro causal links are constituted by micro causal links; and second, it conflicts with physicists' talk about the micro world in causal terms. Therefore the de facto physical asymmetry that we find in the world is not sufficiently general or basic to account for the asymmetry of causation.

However, this argument from microscopic symmetry goes wrong at step two. In fact, because order in our world is not restricted to heat-related phenomena, the notion of entropy is not restricted to macrostates of systems with microstates. Entropy can also be a feature of a purely microscopic system, or of a purely macroscopic system. In fact, the notion of entropy applies to any coarse-grained characterisation of a system.

A purely microscopic example is the case of four molecules in a microscopic box, where each possible combination of positions is equally probable. One relatively ordered state of this system is when the four particles 'form a square,' that is, are positioned on the four corners of any square within the box. The entropy of this state is given in terms of the number of possible combinations that instantiate this formula, relative to the total number of possible combinations of positions. But there's nothing macroscopic about the state 'forms a square.'

A purely macroscopic example is the case of a billiards game. The initial set-up is a low-entropy state in the sense that if we assume that every position on the table is equally probable, for each ball, then the number of possible combinations that constitute a lawful starting configuration is small compared to the total number of possible configurations.

So the same kind of asymmetry that we experience macroscopically can also be seen at a purely microscopic level. Consider, for example, Boltzmann's explanation of macroscopic asymmetry applied to the billiards case. Suppose that when we begin with a highly ordered state (such as the triangular initial configuration) and set the system into motion, then with overwhelming probability the system will evolve towards equilibrium. Until equilibrium is reached, we will witness the kind of asymmetry that we find in our universe; afterwards we will not (unless we first witness a reversed stage where processes move towards higher order). Similarly, if we organise a system of nine O_2 particles into a similar initial configuration, and set it into motion, then with overwhelming probability the system will evolve towards equilibrium; and until equilibrium is reached, we will witness the kind of asymmetry that we find in our universe. Thus the same sort of asymmetry that we experience macroscopically is also visible at a purely microscopic level.

Therefore, if step one of Price's argument is right, and the physical asymmetries (fork, radiation, overdetermination) reduce to entropic asymmetry, then we will find these asymmetries in purely microscopic systems (such as the system of nine O_2 particles) whenever the system is evolving from a highly ordered state to a less ordered state. So, since the microscopic world that constitutes the macroscopic universe we know *is* a system characterised by high initial order, it follows that the kinds of asymmetries appealed to in the various physical theories of causal asymmetry do arise microscopically.[16] It follows, then, that this argument supplies no reason at all to conclude that the fork asymmetry is not a sufficiently widespread and basic phenomenon to constitute the difference between cause and effect.

However, apart from Price's point about the failure of fork asymmetry in the micro world, there is another way that the fork asymmetry might fail to be sufficiently widespread. Suppose a given interaction has a limited number of sets of correlated effects, and that in each case those effects themselves have a common effect, so that in each case we have a closed fork rather than an open fork.[17] Then, since on this theory a process which does *not* part-constitute an open fork has no direction, or its direction is undefined, the theory entails that the processes linking this interaction to its 'effects' have *no* direction.

This problem is particularly acute if, as seems to me to be likely, there aren't that many open forks about. One way out is to hold to the

16. The conflict between reversibility and irreversibility that Boltzmann wrestled with does not necessarily correspond to a macroscopic/microscopic divide, although it does for those kinds of order we call thermodynamic. It does, however, necessarily correspond to a de facto/nomological divide; for in all these cases asymmetry arises from initial conditions, and not from the laws of nature, which are symmetric.
17. Arntzenius (1990) argues in effect that in a deterministic world every conjunctive fork is closed in this way, although his argument requires an unrealistically inclusive account of events.

hope that in fact open forks are prevalent, even if we aren't always aware of them. This hope, which we may dub conjunctivitis, because is envisages hidden conjunctive forks everywhere, may turn out to be little more than wishful thinking. But even if I am wrong about the extent of this problem, the fact remains that there are some actual cases of closed forks, where a process part-constitutes no open fork. It will follow by definition that that causal process has no direction.

VIII.8 FORK THEORY VERSION TWO

Version two, which is in fact the line that Reichenbach took, goes as follows: Processes tend to be linked together in a net of causal processes and interactions. At least some sections of the net constitute open forks. All open forks in fact are open to the future, so the net as a whole can be said to have a direction; namely, the direction of the open forks contained within it. This direction then constitutes the direction of each individual process in the net.[18] If the fork asymmetry thesis has a few exceptions, the theory can be weakened to the effect that the direction of a net is given by the direction of the majority of open forks.[19]

Reichenbach's use of the concept of a net can be spelled out as follows. As we have seen in Chapter 5, there is an 'incoming/outgoing' distinction involved with conservation laws. This is a temporal distinction that is necessary in order for conservation laws to apply to causal interactions. First, the type of temporal concept needed involves a time dimension to give meaning to the idea of conservation, that an object possess the same quantity at different *times*; second, it gives an ordering to allow incoming and outgoing to be distinguished; yet, third, it is isotropic because the conservation laws are symmetric.

This *ordered yet isotropic* notion is the central idea of this section and needs to be elaborated upon. It allows us to distinguish *before* and *after* without preferring one to the other. That is, given an interaction, we need to be able to divide the intersecting causal processes into two classes; we do not give any special significance to one or the other. The

18. See Dowe (1992a) for an extended elaboration of this view.
19. It's important to appreciate that cases of processes converging to or diverging from an interaction are not necessarily cases of statistical forks; they will not be if there is no correlation between the processes. Reichenbach held the view that causality can be reduced to probabilistic relations of this sort; I do not hold that view (see Dowe 1993b).

words 'incoming' and 'outgoing' perhaps carry more import than they ought to for this purpose. They are just labels. This is what is meant by the apparent paradox that we distinguish, yet do not prefer. It remains true that the laws of conservation, and hence definitions CQ1 and CQ2 (see Chapter 5), are symmetric. Thus the various processes involved in a causal interaction can be divided into two classes (incoming and outgoing), but there is as yet no basis for saying that causal influence moves from one to the other.

We make use of the notions of a causal net and causal 'betweenness' due originally to Reichenbach (1956: chap. 5). A causal net (see Figure 8.2) is an interconnected collection of causal interactions (nodes) and causal processes (branches). We will require that the processes be timelike, that is, limited to *either* the backwards or forwards light cones, but not that there be no causal loops. (In that respect the nets defined here are different from Reichenbach's.) An event B is *causally between* events A and C, where A and B are linked by process x, and B and C are linked by process y, if and only if:

1. A, B, C are causal interactions, and x, y are causal processes,
2. x, y are timelike,
3. x, y are not both outgoing and not both incoming with respect to B, and
4. no conservation law is violated at B.

This definition clearly rests on the possibility of distinguishing 'incoming' from 'outgoing' as discussed earlier. Notice that this betweenness relation is symmetric with respect to the outer interactions, but not otherwise. If B is causally between A and C, then it is also true that B is causally between C and A. Thus it has a type of order that has no preferred direction, which reflects the distinction between incoming and outgoing processes. The causal betweenness relation will apply to all connected processes and interactions within a net except where a causal interaction is not governed by conservation laws.

We can also define a concept of *alignment* : If B is causally between A and C, then x and y exhibit a relation of 'parallel alignment' or just 'alignment' (where x is a process linking A and B, and y is a process linking B and C). That is, 'x and y are aligned' indicates that x and y have the same otherwise indeterminate direction. They point the same way, but we do not know which way. As was pointed out earlier, this distinction is necessary for conservation laws to be applied to any inter-

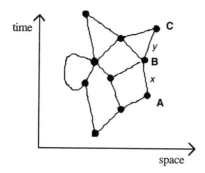

Figure 8.2. A causal net.

action. If all the processes that intersect in a causal interaction pointed toward or all pointed away from the interaction, then conservation would be violated. This idea of alignment replaces Reichenbach's concept of the local comparability of time order (1956: 34–35), which Grünbaum has exposed as inadequate (1973: 193, note 2) on the grounds that it is epistemically dependent. If all the processes in a causal net are timelike and there are no causal loops, then it follows that the net has a *global alignment*, meaning that all the processes in the net point the same way, although we do not know which way. However, we have not ruled out the possibility that processes in a causal net form causal loops. So suppose we take it as a contingent fact that causal loops are either nonexistent or very rare. Then causal nets in the real world have a global alignment, meaning that most processes point the same way, but we do not know which way. This roughly corresponds to Reichenbach's concept of 'lineal order.'

In summary, causal nets display this distinction between order and direction: causal nets are symmetric, they do not have a preferred orientation or direction, but they do have a global alignment. Reichenbach's (modified) argument is this: if a causal net has a global alignment, then if any part or aspect of the net has a direction, then every process has that direction. Causal nets by themselves do not provide the sort of analysis of causal asymmetry that we seek, but Reichenbach's idea was that the direction of open forks in the net defines the direction of the net. So, in turn, the direction of each process in the net is also defined.

Perhaps more needs to be said about the extent of the net. Suppose the causal net in which a process is found extends a great distance in space and time. In this case, is the 'net' the entire net, or just a localised region of this net? We should take the answer to be the entire net, extending to the entire universe for substantial periods of time, applying if necessary to that spacetime region to which the phenomenological Second Law of Thermodynamics applies. (If it extends any further, say to the heat death of the universe, then there will be no direction to which the majority of open forks point.)

This second version of the fork theory has, as an advantage over the first, the consequence that all processes have a direction. Any process that is part of a wider net of causal processes and interactions takes on the direction of the net as a whole. If there are isolated nets, which are not connected to each other in any way, then those nets may have different directions, but still, every process in those nets will have a definite direction. The only cases that will fail to have a definite direction will be cases where there is a net containing no open forks, or no clear majority of open forks in either direction, such as in the heat death of the universe.

But, while the second version avoids that problem, it also faces a more serious objection: it is incompatible with the backwards causation model. Consider again the process between S (the source) and M_a (a measurement) in the Bell set-up. On this theory, the direction of this process is determined by the causal net in which it is found, which in turn is determined by the direction of the open forks in that net. Now, a particular process cannot have a direction counter to those of the surrounding processes to which it is connected. But that is exactly what is proposed in the backwards causality model. The direction of causal influence from M_a-S is postulated to be the reverse of the direction of the influence from S-M_b, and the two processes must be connected together in a net. An entire net, unconnected to the rest of the universe, could take on a counterdirection (such as may be the case with antimatter universes), but that is not the scenario in the Bell set-up. It is analytic that individual processes within a net cannot take different directions, if version two is correct. So Reichenbach's version rules out the kind of backwards causation required for the backwards causation model of Bell phenomena.[20]

20. This would not particularly worry Reichenbach. In fact, he went to some lengths to prove that such "anomalies," as he called them, don't occur. Reichenbach's purpose in proposing a fork theory of the direction of causation was the devel-

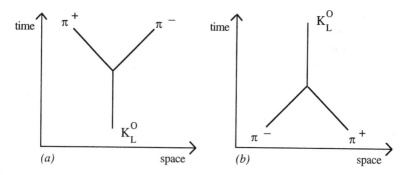

Figure 8.3. (a) kaon decay. (b) T reversal.

VIII.9 THE KAON THEORY

Another version of the physical approach to the direction of causation, the kaon theory, also faces this kind of difficulty. The *kaon arrow* was first postulated in 1964 when it was discovered that the K_L^0 meson[21] can decay into two π mesons, which violates the principle of CP invariance.[22] Since the charge C and parity P symmetries are broken, it follows from fundamental considerations of relativity theory (Lee et al. 1957) that T invariance is also violated. This means that the laws of physics are not completely time-symmetric and that K meson decay is not completely time-reversible. To spell out what this means consider an example (see Figure 8.3):

$$K_L^0 \rightarrow \pi^+ + \pi^-$$

If we consider the reversed movie track of this interaction, it will look like Figure 8.3(b). To test the T reversibility of the process in Figure 8.3(a), one has to produce the interaction in Figure 8.3(b) experimentally. If the interaction in (b) occurs at a different rate than the interaction in (a), then T invariance is violated. It is not necessarily an easy task to produce the right boundary conditions to do this (Sachs 1972: 595). However, experimental work on K meson decay patterns over the

opment of a causal theory of time. This requires that the causal direction not be defined in terms of time; but the causal theory of time itself rules out backwards-in-time causation.

21. The L refers to relatively long-lived (that is, about 5×10^{-8} sec) neutral K mesons (Sachs 1987: 193).

22. For a full account, see Sachs (1987: chap. 9).

last thirty years has produced conclusive evidence of CP violation (see Sachs 1987: chap. 9).

What such a process represents, then, is a process oriented in time, that is, asymmetric with respect to time. Unlike the entropic arrow, this physical asymmetry (which we will refer to as the kaon arrow) is nomological, not de facto, and local, not global. We wish to consider, then, whether the kaon arrow can provide the CQ theory with a physical basis for causal asymmetry. First we need to recognise that such symmetry-breaking processes are very rare. The violations occur less than one percent of the time (Layzer 1975: 58), a fact that physicist Robert Sachs describes as a great cause for mystification because "the breaking of a symmetry should occur with a resounding roar rather than with a whimper" (Sachs 1972: 595). Also, K mesons only appear in high-energy experiments, and they are very short-lived, so you never come across them in everyday life (Layzer 1975). Many philosophers therefore draw the conclusion that the kaon arrow can have no relevance to macrofeatures of the world. (See, for example, Layzer 1975: 58; Mehlberg 1980: 183; Skyrms 1980: 120; Swinburne 1981: 229; and Whitrow 1980: 336.) However, physicists have not been so quick to draw this conclusion given the uncertainty that exists concerning the relations among the various so-called arrows of time, referring here not only to the entropic and kaon arrows but also the cosmological, electromagnetic and subjective arrows (see Cramer 1986; 1988a). Indeed, some physicists have suggested a link between the kaon arrow and the cosmological arrow on account of the likely action of CP violation in the Big Bang and the dominance of matter over antimatter in the universe (Cramer 1988b: 1207). However, it is not my purpose to pursue this possibility here.

The purpose, rather, is to see in what sense the kaon arrow can be used as the physical basis for causal asymmetry. As in the case of the second version of the fork theory, we make use of the idea of a causal net with a global alignment. We construct a causal net that contains both the process in question and a T-violating kaon decay, which can be done in practice by the use of a clock. The orientation of the kaon decay defines a direction for all processes in the net, since it is aligned, and hence defines a direction for the causal process in question. Thus kaon decay operates as something of a direction marker. Therefore we can formulate a kaon definition of causal asymmetry:

The direction of a causal process is defined by the direction of T-violating kaon processes.

One advantage of the kaon theory over the fork theory is its nomo-logical character, at least for those who would demand this of a theory of causality. A disadvantage is that the kaon arrow lacks the all-pervasive, ubiquitous nature of the entropic arrow.

The kaon theory was presented by myself in Dowe (1992a), where it was argued that this account is compatible with the backwards cau-sation model. But as is clear from the earlier discussion, this was a mistake. The kaon theory clearly suffers from the same difficulty as the second version of the fork theory, because it too takes the direction of a particular causal process to be constituted by the direction of the net in which the process is located.

On this score, version one of the fork theory is clearly preferable to either the second version or the kaon theory. If the direction of a causal process is given by the direction of an open conjunctive fork part-constituted by that process, then, if the asymmetry thesis is universally true, there is no backwards causation. But if the asymmetry thesis admits of exceptions – occasional cases where forks are open to the past – then backwards causation is not only possible but actual: the processes constituting those forks have a backwards direction.[23] This opens up the possibility that the backwards model of Bell systems will work, although only if the process from M_a to S part-constitutes a con-junctive fork open to the past – a point to which we will return.

So, to summarise the situation so far, version one of the fork theory faces the problem that many processes will most probably not have a direction, but it does leave open the possibility of backwards causation; while version two of the fork theory and the kaon theory both bestow a direction on all processes but rule out backwards causation. Fortu-nately, a third version achieves a synthesis wherein both problems are solved.

VIII.10 FORK THEORY VERSION THREE

Version three goes as follows: The direction of a causal process is given by the direction of an open conjunctive fork part-constituted by that process; or, if there is no such conjunctive fork, by the direction of the majority of open forks contained in the net in which the process is

23. Providing that certain problems of indeterminacy and ambiguity can be avoided; for example, where the process part-constitutes a fork open to the future and also part-constitutes a fork open to the past.

found. This account simply disjoins the conditions of version one and version two in such a way as to give primacy to version one. The clause from version two is, as it were, the back-up measure. Consequently, there is a sense in which there are two grades, or ways in which a process can have a direction. The primary kind of direction derives from direct participation in an open fork; the secondary kind derives indirectly, by association, from the net in which the process is located. Whatever the shortcomings of this suggestion may be, it solves the problems that the other two versions face: it does bestow a direction on all processes, and it does leave open the possibility of backwards causation.

On this account (and also for fork version 1), a causal process broadly speaking (that is, in the sense of Chapter 5) can have two directions. The reason is that the fork account refers to conjunctive forks, which concern the frequencies of event-types. On the process approach, as developed in Chapter 7, such events need to be reducible to objects possessing quantities. So, since an object can possess more than one quantity at a time, it may well be that simultaneous events concerning the same object at a single time enter into quite different conjunctive forks. For example, a measurement interaction M and the earlier preparation state S of an object may be connected as cause and effect, and as effect and cause. Huw Price, for example (1996a), regards this as a counterintuitive consequence, but in fact it is just as well, because the backwards models of quantum interactions, such as that of Cramer, involve exactly this dual direction. Causal influence goes in both directions in such quantum events.

There is another objection that needs to be mentioned here. According to an objection due to Michael Tooley, the fork theory has the consequence that the direction of a causal process is made true by things *extrinsic* to the process. This is counterintuitive, Tooley says, especially if we are faced with a situation where the direction of a particular process is determined by correlated effects that occur much later (Tooley 1987: 237; 1993: 22). This objection applies to the different versions to differing extents. According to version one, the direction of a process is given by the direction of the open fork it part-constitutes, which is a relatively local affair, compared to version two, according to which the direction is given by the net in which it is located. Version three shares this defect, if it is a defect.

But Tooley's principle that the direction of a causal process is something intrinsic to a process is not something that can be proved simply by appeal to such an intuition. After all, that intuition may be false, or at least, I don't think it is something that the proponent of the fork asymmetry theory need bow to, especially in the absence of any satisfactory account that tells us what the *intrinsic* direction of a causal process amounts to. Consider the cases of colour and heat. At one time, folk intuitions about colour would have said that colour was intrinsic to the objects; that proved false. Or, even closer to home, folk intuitions may tell us that certain irreversible processes have an intrinsic direction, which we also know to be false. So, I suggest that we simply accept this consequence and admit that we must appeal to extrinsic features, and that this may well contradict folk intuitions.[24]

It remains, then, to outline the promised empirical consequence of the fork theory, which, if proved correct, would constitute confirmation of the backwards causation model of Bell phenomena. I will then conclude with some comments about the implications of rejecting the fork theory.

VIII.11 THE EMPIRICAL PREDICTION

If version one or three of the fork theory of the direction of causal processes is correct, then cases of backwards causation involve conjunctive forks open to the past. In the case of Bell phenomena, this means that the backwards process M_a-S is accompanied by another backwards process M_a-x, where x is an unknown factor correlated with S such that S, M_a and x form a conjunctive fork. The unknown factor x will be some factor in the apparent causal past of the measurement, perhaps something in the past of the measuring device or the experimenter, if one is directly involved. For example, it may be an itch that the experimenter feels about the time the experiment begins. If so, then the itch x and separating of the particles S will be correlated, such that

$$P(S.x) > P(S) P(x)$$

---

24. It's worth just noting at this point that *the* most famous account of causation, the Humean regularity account, has the consequence that whether a process is causal or not depends on what happens in remote areas of our spacetime universe. Yet numerous philosophers, including many who thought the purpose of philosophy consists in elucidating the structure of our commonsense concepts, have held this position. This must count as prima facie evidence that intrinsically is not essential to the folk concept.

This is something that in principle could be sought for, and found, experimentally. It is also necessary that there be no earlier event C such that C screens off the dependence between x and S; if this is the case, then we have empirically confirmed the case of backwards causation. As an empirical prediction we can only provide here a most abstract characterisation of the proposal. The general form of such a test would be to examine Bell experiments looking for factors that have not previously been thought relevant, to see whether we can discover a previously unnoticed correlation between the separating of the particles and some factor in the 'causal past' of the measurement.

There is no reason why S need always to be correlated with the same kind of event x. It just needs to be correlated with some event x, y, or z. This would depend, presumably, on the experimental arrangement. Just as with the firing of a gun, where the powder on my hand will be correlated with some kind of destruction somewhere, the kind depending on the circumstances, so also S will be correlated with some kind of event in the past of the measurement, the kind depending on the circumstances.[25]

No doubt this implication will seem implausible to many. Hausman says as much, in fact (Hausman 1998: 251). But to reject this implication will entail either rejecting the reasoning, or one of the premises. Since Hausman accepts the premises that backwards causation is possible and that some form of the fork theory is correct, rejection of the implication leaves him with the challenge of saying what is wrong with the reasoning.

It may be thought that the existence of such a correlation between x and S would close the bilking loophole, because one could know that x had occurred, and then prevent M_a (its cause).[26] However, this is not so. To know that x is an effect of M_a one would have to know not only that x occurred, but also that S occurred (and by S we mean the state including the hidden part of S, as caused by M_a). But we cannot know that S has occurred until M_a and R_b occur. And so we cannot know that

---

25. Perhaps one difficulty this suggestion will face in achieving any level of acceptance will be simply that the unknown factors seem so irrelevant. What is relevant in experimentation is for the most part set by theoretical considerations, and major theoretical developments tend to change our view of what is relevant. In our case, admitting backwards causation is a major conceptual change, and therefore can be expected to change our view of what is relevant.

26. This difficulty was pointed out to me by Nick Smith.

x is the effect of M_a until M_a and R_b have occurred, and so bilking is not physically possible, even with knowledge of x.

VIII.12 SOME IMPLICATIONS OF REJECTING THE FORK THEORY

To finish, it's worth considering briefly what conclusions we would be forced to if the backwards causality model is correct, but contrary to our earlier arguments, the fork theory is in fact unsatisfactory. For example, suppose that one objects to the view that the direction of a causal process is extrinsic to the process, to the view that it is a de facto matter, or to its disjunctive nature. Then, since we started with the premise of backwards causation, which requires a notion of a direction of a causal process, we still owed an account of that direction. Our arguments against the temporal and the subjective accounts are comprehensive in the sense that they rule out any kind of temporal or subjective theory. But in the case of the physical theory, we have simply ruled out the extant versions of such a theory. We would conclude from this that there must be a satisfactory physical theory of a different kind from the ones we have considered, which perhaps no one has yet thought of. Such a theory would identify the intrinsic and/or nomological feature of a causal process that constitutes its direction.

Such a view entails a very much more radical revision of current science than is envisaged in the empirical prediction outlined here. It would entail that the current laws of physics are radically incomplete, for this new physical theory would identify laws of nature that are in some way not time-symmetric.[27] This is radical indeed, and makes my suggestion look fairly ordinary. This, I imagine, simply provides all the more reason to accept my conclusion.

For reasons similar to the sorts of reasons listed earlier, Tooley proposes what he calls a 'realist' account of the direction of causation (1987). Now, a realist approach prompts the following question: Is the direction of physical causal processes itself something essentially physical? (It seems that Tooley says no, since he holds that direction is not reducible to physics.) If the answer is yes, then we should follow the earlier reasoning and infer that current physics is radically incomplete. If the answer is no, and the direction of causation is not something essentially physical, then the realist account is open to three objections.

---

27. It would identify features that are much more pervasive than kaon decay, the only known case of T-symmetry violation.

First, it seems to relegate the direction of a causal process to the realm of the epistemically inaccessible.[28] Second, even if we can get past the first point, we still need to ask why a no answer would be preferable to a yes answer. To prefer to say that something is in principle irreducible to physics, rather than as yet not reduced, depends on both an unfounded pessimism about future physics and an unfounded belief in the completeness of current physics. Third, it's not clear that Tooley's view is compatible with the backwards causality model, since that model seems to reveal to us pointers to a physical direction.

Thus it seems to me that there is a very strong case for saying that the backwards causality model of Bell phenomena leads us to a fork account of the direction of causal processes.

VIII.13 SUMMARY

In this final chapter I have argued for a version of the fork asymmetry account of the direction of causal processes. This argument has proceeded from a controversial premise, viz., the truth of the backwards-in-time causation model of Bell phenomena in quantum physics. However, there is some hope that further evidence for that model could arise from the empirical prediction deduced here from the conjunction of that physical model and the philosophical theory of causal asymmetry.

This premise, the backwards causation model, rules out many of the extant theories of causal asymmetry, including the temporal, subjective and kaon theories. It also rules out one version of the fork theory, that due to Reichenbach, where the direction of a process is determined by the direction of the net in which it is found. Another version of the fork theory, which requires that a process part-constitute an open conjunctive fork, was ruled out for a different reason, viz., it leaves too many processes with no defined direction. However, a compromise version was found to avoid both these difficulties. According to the theory argued for here, the direction of a causal process is given by the direction of an open conjunctive fork part-constituted by that process; or, if there is no such conjunctive fork, by the direction of the majority of open forks contained in the net in which the process is found. This

---

28. That is, Tooley's account seems to me to be open to the kinds of objections developed by Van Fraassen (1989) against Armstrong's realism about physical necessitation.

allows backwards causation where an open fork points to the past, given that the majority of forks point to the future. It also defines a direction for all processes.

This provides an answer to, and a motivation for, our third and final question, *what makes a cause different from an effect?*, in a way that follows naturally from the account given in Chapters 5–7.

References

Ackerman, F. 1995: "Analysis", in Kim, J. and Sosa, E. (eds.), *A Companion to Metaphysics*, Oxford: Blackwell, 9–11.

Alston, W. 1967: "Philosophy of Language", in Edwards, P. (ed.), *Encyclopedia of Philosophy*, New York: Macmillan, 386–390.

Anscombe, E. 1971: *Causality and Determinism*. Cambridge: Cambridge University Press.

Armstrong, D. 1997: *A World of States of Affairs*. Cambridge: Cambridge University Press.

Armstrong, D. 1999: "The Open Door", in Sankey, H. (ed.), *Causation and Laws of Nature*, Dordrecht: Kluwer.

Armstrong, D. 1978: *Nominalism and Realism*. Cambridge: Cambridge University Press.

Armstrong, D. 1980: "Identity through Time", in van Inwagen, P. (ed.), *Time and Cause*, Dordrecht: Reidel, 67–78.

Armstrong, D. 1983: *What Is a Law of Nature?* Cambridge: Cambridge University Press.

Arntzenius, F. 1990: "Physics and Common Causes", *Synthese* 82: 77–96.

Aronson, J. 1971a: "The Legacy of Hume's Analysis of Causation", *Studies in History and Philosophy of Science* 2: 135–156.

Aronson, J. 1971b: "On the Grammar of 'Cause'", *Synthese* 22: 414–430.

Aronson, J. 1982: "Untangling Ontology from Epistemology in Causation", *Erkenntnis* 18: 293–305.

Aronson, J. 1985: "Conditions versus Transference: A Reply to Ehring", *Synthese* 63: 249–257.

Arya, A. 1974: *Elementary Modern Physics*. Reading, Mass.: Addison-Wesley.

Barker, S. 1998: "Predetermination and Tense Probabilism", *Analysis* 58: 290–296.

Beauchamp, T. and Rosenberg, A. 1981: *Hume and the Problem of Causation*. New York: Oxford University Press.

Belinfante, F. 1973: *A Survey of Hidden-Variables Theories*. Oxford: Pergamon Press.

Bell, J. 1964: "On the Einstein Podolsky Rosen Paradox", *Physics* 1: 195–200.

Bennett, J. 1984: *A Study of Spinoza's Ethics*. Indianapolis: Hackett.

Bennett, J. 1995: *The Act Itself*. New York: Clarendon.

Bigelow, J., Ellis, B., and Pargetter, R. 1988: "Forces", *Philosophy of Science* 55: 614–630.

Bigelow, J. and Pargetter, R. 1990a: "Metaphysics of Causation", *Erkenntnis* 33: 89–119.

Bigelow, J. and Pargetter, R. 1990b: *Science and Necessity*. Cambridge: Cambridge University Press.

Bjelke, E. 1975: "Dietary Vitamin A and Human Lung Cancer", *International Journal of Cancer* 15: 561–565.

Blackburn, S. 1990: "Hume and Thick Connections", *Philosophy and Phenomenological Research* 50: 237–250.

Broughton, J. 1987: "Hume's Ideas about Necessary Connection", *Hume Studies* 13: 217–244.

Carrier, M. 1998: "Salmon 1 versus Salmon 2: Das Prozeßmodell der Kausalität in seiner Entwicklung (The Process Model of Causality in Its Development)", *Dialektik* 49–70.

Cartwright, N. 1979: "Causal Laws and Effective Strategies", *Nous* 13: 419–437.

Cartwright, N. 1983: *How the Laws of Physics Lie*. Oxford: Clarendon.

Collingwood, R. 1974: "Three Senses of the Word 'Cause'", in Beauchamp, T. (ed.), *Philosophical Problems of Causation*, Encino, Calif.: Dickenson, 118–126.

Costa, M. 1989: "Hume and Causal Realism", *Australasian Journal of Philosophy* 67: 172–190.

Craig, E. 1987: *The Mind of God and the Works of Man*. Oxford: Clarendon.

Cramer, J. 1986: "The Transactional Interpretation of Quantum Mechanics", *Reviews of Modern Physics* 58: 647–687.

Cramer, J. 1988a: "An Overview of the Transactional Interpretation of Quantum Mechanics", *International Journal of Theoretical Physics* 27: 227–236.

Cramer, J. 1988b: "Velocity Reversal and the Arrows of Time", *Foundations of Physics* 18: 1205–1212.

Davidson, D. 1969: "The Individuation of Events", in Rescher, N. (ed.), *Essays in Honour of Carl G. Hempel*, Dordrecht: Reidel, 216–234.

Davis, W. 1988: "Probabilistic Theories of Causation", in Fetzer, J. (ed.), *Probability and Causality: Essays in Honor of Wesley C. Salmon*, Dordrecht: Reidel, 133–160.

d'Espagnat, B. 1979: "The Quantum Theory and Reality", *Scientific American* 241: 128–140.

Dieks, D. 1981: "A Note on Causation and the Flow of Energy", *Erkenntnis* 16: 103–108.

Dieks, D. 1986: "Physics and the Direction of Causation", *Erkenntnis* 25: 85–110.

Dowe, P. 1989: "On Tooley On Salmon", *Australasian Journal of Philosophy* 64: 469–471.

Dowe, P. 1990: "Probabilistic Theories of Causation," Master's thesis, Macquarie University.

Dowe, P. 1992a: "An Empiricist Defence of the Causal Account of Explanation", *International Studies in the Philosophy of Science* 6: 123–128.

Dowe, P. 1992b: "Process Causality and Asymmetry", *Erkenntnis* 37: 179–196.

Dowe, P. 1992c: "Wesley Salmon's Process Theory of Causality and the Conserved Quantity Theory", *Philosophy of Science* 59: 195–216.

Dowe, P. 1993a: "The Anti-realism of Costa de Beauregard", *Foundations of Physics Letters* 6: 469–475.

Dowe, P. 1993b: "On the Reduction of Process Causality to Statistical Relations", *British Journal for the Philosophy of Science* 44: 325–327.

Dowe, P. 1996: "J. J. C. Smart and the Rise of Scientific Realism", in Dowe, P., Nicholls, M. and Shotton, L. (eds.), *Australian Philosophers*, Hobart: Pyrrho Press, 25–37.

Dowe, P. 1997a: "A Defense of Backwards-in-time Causation Models in Quantum Mechanics", *Synthese* 112: 233–246.

Dowe, P. 1997b: "Re-reading Russell," conference paper read at the Australian Association of Philosophy conference, July.

Dowe, P. 1999: "Good Connections: Causation and Causal Processes", in Sankey, H. (ed.), *Causation and Laws of Nature*, Dordrecht: Kluwer, 323–346.

Dowe, P. unpublished-a: "A Dilemma for Objective Chance," conference paper read at the Australian Association of Philosophy conference, July, 1992.

Dowe, P. unpublished-b: "Singular Causation," conference paper read at the Australian Association of Philosophy conference, University of Adelaide, July, 1993.

Dowe, P. unpublished ms: "Defending Time Travel."

Ducasse, C. 1926: "On the Nature and the Observability of the Causal Relation", *Journal of Philosophy* 23: 57–68.

Ducasse, C. 1969: *Causation and the Types of Necessity*. New York: Dover.

Ducasse, C. 1976: "Causality: Critique of Hume's Analysis", in Brand, M. (ed.), *The Nature of Causation*, Urbana: University of Illinois Press, 67–76.

Dummett, M. 1964: "Bringing about the Past", *Philosophical Review* 73: 338–359.

Earman, J. 1976: "Causation: A Matter of Life and Death", *The Journal of Philosophy* 73: 5–25.

Earman, J. 1986: *A Primer on Determinism*. Dordrecht: Reidel.

Eells, E. 1988: "Probabilistic Causal Levels", in Skyrms, B. and Harper, W. (eds.), *Causation, Chance and Credence Volume 2*, Dordrecht: Kluwer, 109–133.

Eells, E. 1991: *Probabilistic Causality*. Cambridge: Cambridge University Press.

Eells, E. and Sober, E. 1983: "Probabilistic Causality and the Question of Transitivity", *Philosophy of Science* 50: 35–57.

Ehring, D. 1986: "The Transference Theory of Causation", *Synthese* 67: 249–258.

Ehring, D. 1991: "Motion, Causation, and the Causal Theory of Identity", *Australasian Journal of Philosophy* 69: 180–194.

Ehring, D. 1998: *Causation and Persistence*. Oxford: Oxford University Press.

Emmet, D. 1985: *The Effectiveness of Causes*. Albany: State University of New York Press.

Enge, H. 1966: *Nuclear Physics*. Amsterdam: Addison.

Fair, D. 1979: "Causation and the Flow of Energy", *Erkenntnis* 14: 219–250.

Fetzer, J. 1987: "Critical Notice: Wesley Salmon's Scientific Explanation and the Causal Structure of the World", *Philosophy of Science* 54: 597–610.

Fetzer, J. 1988: "Probabilistic Metaphysics", in Fetzer, J. (ed.), *Probability and Causality*, Dordrecht: Reidel, 109–132.

Forge, J. 1982: "Physical Explanation: With Reference to the Theories of Scientific Explanation of Hempel and Salmon", in McLaughlin, R. (ed.), *What? Where? When? Why?*, Dordrecht: Reidel, 211–219.

Forge, J. 1985: "Book Review: Salmon, *Scientific Explanation and the Causal Structure of the World*", *Australasian Journal of Philosophy* 63: 546–585.

Ganeri, J., Noordhof, P., and Ramachandran, M. 1996: "Counterfactuals and Preemptive Causation", *Analysis* 55: 219–225.

Garrett, D. 1993: "The Representation of Causation and Hume's Two Definitions of 'Cause' ", *Nous* 27: 167–190.

Gasking, D. 1955: "Causation and Recipes", *Mind* 64: 479–487.

Gasking, D. 1996: *Language, Logic and Causation*. ed. Oakley, I. and O'Neill, L. Melbourne: Melbourne University Press.

Giere, R. 1988: "Book Review: Salmon, *Scientific Explanation and the Causal Structure of the World*", *The Philosophical Review* 97: 444–446.

Glover, J. 1977: *Causing Death and Saving Lives*. Harmondsworth: Penguin.

Goldman, A. I. 1977: "Perceptual Objects", *Synthese* 35: 257–284.

Good, I. 1961: "A Causal Calculus – 1", *British Journal for the Philosophy of Science* 11: 305–318.

Good, I. 1962: "A Causal Calculus – 2", *British Journal for the Philosophy of Science* 12: 43–51.

Goodman, N. 1955: *Fact, Fiction and Forecast*. Cambridge: Harvard University Press.

Grünbaum, A. 1973: *Philosophical Problems of Space and Time*. 2nd ed., Dordrecht: Reidel.

Hall, N. unpublished: "Two Concepts of Causation."

Hanna, J. 1986: "Book review: Salmon, *Scientific Explanation and the Causal Structure of the World*", *Review of Metaphysics* 39: 582.

Harré, R. 1985: "Book Review: A Realist Philosophy of Science", *Philosophy of Science* 52: 483–485.

Hart, H. and Honore, T. 1985: *Causation in the Law*. Oxford: Clarendon.

Hausman, D. 1984: "Causal Priority", *Nous* 18: 261–279.

Hausman, D. 1998: *Causal Asymmetries*. New York: Cambridge University Press.

Heathcote, A. 1989: "A Theory of Causality: Causality = Interaction (as Defined by a Suitable Quantum Field Theory)", *Erkenntnis* 31: 77–108.

Hesslow, G. 1976: "Two Notes on the Probabilistic Approach to Causality", *Philosophy of Science* 43: 290–292.

Hirsch, E. 1982: *The Concept of Identity*. New York: Oxford University Press.

Hitchcock, C. 1995: "Salmon on Explanatory Relevance", *Philosophy of Science* 62: 304–320.

Horwitz, L. P., Arshansky, R. I., and Elitzur, A. C. 1988: "On the Two Aspects of Time: The Distinction and Its Implications", *Foundations of Physics* 18: 1159–1194.

Hospers, J. 1990: *An Introduction to Philosophical Analysis*. 3rd ed., London: Routledge.

Hume, D. 1975: *An Enquiry Concerning Human Understanding*, ed. Selby-Bigge, L. A. and Nidditch, P. H. 3rd ed., Oxford: Clarendon.

Hume, D. 1978: *A Treatise of Human Nature*, ed. Selby-Bigge, L. A. and Nidditch, P. H. 2nd ed., Oxford: Clarendon.

Humphreys, P. 1981: "Aleatory Explanations", *Synthese* 48: 225–232.

Humphreys, P. 1985: "Why Propensities Cannot be Probabilities", *The Philosophical Review* 94: 557–570.

Humphreys, P. 1986: "Book Review: Salmon, *Scientific Explanation and the Causal Structure of the World*", *Foundations of Physics* 16: 1211–1216.

Humphreys, P. 1989: *The Chances of Explanation*. Princeton: Princeton University Press.

Jackson, F. 1992: "Critical Notice of S. Hurley's *Natural Reasons*", *Australasian Journal of Philosophy* 70: 475–488.

Jackson, F. 1994: "Armchair Metaphysics", in Michael, M. and O'Leary-Hawthorne, J. (eds.), *Philosophy in Mind*, Dordrecht: Kluwer, 23–42.

Kim, J. 1969: "Events and Their Descriptions: Some Considerations", in Rescher, N. (ed.), *Essays in Honour of Carl G. Hempel*, Dordrecht: Reidel, 198–215.

Kistler, M. 1998: "Reducing Causality to Transmission", *Erkenntnis* 48: 1–24.

Kitcher, P. 1985: "Two Approaches to Explanation", *The Journal of Philosophy* 82: 632–639.

Kitcher, P. 1989: "Explanatory Unification and the Causal Structure of the World", in Kitcher, P. and Salmon, W. (eds.), *Minnesota Studies in the Philosophy of Science Volume XIII*, Minneapolis: University of Minnesota Press, 410–505.

Kolak, D. and Martin, R. 1987: "Personal Identity and Causality: Becoming Unglued", *American Philosophical Quarterly* 24: 339–347.

Kripke, S. 1980: *Naming and Necessity*. Oxford: Blackwell.

Kvart, I. 1997: "Cause and Some Positive Causal Impact", in Tomberlin, J. (ed.), *Philosophical Perspectives, 11; Mind, Causation, and World*, Boston: Blackwell, 401–432.

Layzer, D. 1975: "The Arrow of Time", *Scientific American* 233: 56–69.

Lee, T., Oehme, R., and Yang, C. N. 1957: "Remarks on Possible Noninvariance under Time Reversal and Charge Conjugation", *Physical Review* 106: 340–345.

Leslie, J. 1986: "The Scientific Weight of Anthropic and Teleological Principles", in Rescher, N. (ed.), *Current Issues in Teleology*, Lantham: University Press of America, 111–119.

Lewis, D. 1983: *Philosophical Papers Volume I*. Cambridge: Cambridge University Press.

Lewis, D. 1986: *Philosophical Papers Volume II*. New York: Oxford University Press.

Lewis, D. 1994: "Reduction of Mind", in Guttenplan, S. (ed.), *A Companion to Philosophy of Mind*, Cambridge, Mass.: Blackwell, 412–431.

Mackie, J. 1974: *The Cement of the Universe*. Oxford: Clarendon.

Mackie, J. 1985: *Logic and Knowledge – Selected Papers*. Oxford: Clarendon.

Mackie, J. 1992: "Causing, Delaying and Hastening: Do Rains Cause Fires?" *Mind* 101: 483–500.

Mehlberg, H. 1980: *Time, Causality, and the Quantum Theory Volume 2*. Dordrecht: Reidel.

Mellor, D. 1971: *The Matter of Chance*. Cambridge: Cambridge University Press.

Mellor, D. 1988: "On Raising the Chances of Effects", in Fetzer, J. (ed.), *Probability and Causality: Essays in Honour of Wesley C. Salmon*, Dordrecht: Reidel, 229–239.

Mellor, D. 1995: *The Facts of Causation*. London: Routledge.

Menzies, P. 1989a: "Probabilistic Causation and Causal Processes: A Critique of Lewis", *Philosophy of Science* 56: 642–663.

Menzies, P. 1989b: "A Unified Account of the Causal Relata", *Australasian Journal of Philosophy* 67: 59–83.

Menzies, P. 1996: "Probabilistic Causation and the Pre-emption Problem", *Mind* 105: 85–117.

Menzies, P. and Price, H. 1993: "Causation as a Secondary Quality", *British Journal for the Philosophy of Science* 44: 187–204.

Millikan, R. 1989: "In Defense of Proper Functions", *Philosophy of Science* 56: 288–302.

Mullany, N. 1992: "Common Sense Causation – An Australian View", *Oxford Journal of Legal Studies* 12: 421–439.

Musgrave, A. 1977: "Explanation, Description and Scientific Realism", *Scientia* 112: 727–742.

Neander, K. 1991: "Functions as Selected Effects: The Conceptual Analyst's Defense", *Philosophy of Science* 58: 168–184.

Noordhof, P. 1999: "Probabilistic Causation, Preemption and Counterfactuals", *Mind* 108: 95–126.

Nozick, R. 1981: *Philosophical Explanations*. Cambridge, Mass.: Harvard University Press.

O'Leary-Hawthorne, J. and Price, H. 1996: "How to Stand Up for Non-Cognitivists", *Australasian Journal of Philosophy* 74: 275–292.

Papineau, D. 1989: "Pure, Mixed and Spurious Probabilities and Their Significance for a Reductionist Theory of Causation", in Kitcher, P. and Salmon, W. (eds.), *Minnesota Studies in the Philosophy of Science XIII*, Minneapolis: University of Minnesota Press, 410–505.

Papineau, D. 1993: "Can We Reduce Causal Direction to Probabilities", in Hull, D., Forbes, M., and Okruhlik, K. (eds.), *PSA 1992 Volume 2*, East Lansing, Mich.: Philosophy of Science Association, 238–252.

Papineau, D. and Sober, E. 1986: "Causal Factors, Causal Inference, Causal Explanation", *Proceedings of the Aristotelian Society Supplementary* 60: 115–136.

Parfit, D. 1984: *Reasons and Persons*. Oxford: Clarendon.

Pears, D. 1990: *Hume's System*. Oxford: Oxford University Press.

Pitt, J., ed. 1988: *Theories of Explanation*. Oxford: Oxford University Press.

Pitt, V., ed. 1977: *The Penguin Dictionary of Physics*. Harmondsworth: Penguin.

Price, H. 1991: "Agency and Probabilistic Causality", *British Journal for the Philosophy of Science* 42: 157–176.

Price, H. 1992: "Agency and Causal Asymmetry", *Mind* 101: 501–520.

Price, H. 1993: "The Direction of Causation: Ramsey's Ultimate Contingency", in Hull, D., Forbes, M., and Okruhlik, K. (eds.), *PSA 1992 Volume 2*, East Lansing, Mich.: Philosophy of Science Association, 253–267.

Price, H. 1994: "A Neglected Route to Realism about Quantum Mechanics", *Mind* 103: 303–336.

Price, H. 1996a: "Backwards Causation and the Direction of Causal Processes: Reply to Dowe", *Mind* 105: 467–474.

Price, H. 1996b: *Time's Arrow and Archimedes' Point*. Oxford: Oxford University Press.

Quine, W. 1965: *The Ways of Paradox*. New York: Random House.

Quine, W. 1973: *The Roots of Reference*. La Salle, Ill.: Open Court.

Ramachandran, M. 1997: "A Counterfactual Analysis of Causation", *Mind* 106: 264–277.

Reichenbach, H. 1956: *The Direction of Time*. Berkeley: University of California Press.

Reichenbach, H. 1958: *The Philosophy of Space and Time*. New York: Dover.

Reichenbach, H. 1991: *The Direction of Time*. 2nd ed., Berkeley: University of California Press.

Reuger, A. 1998: "Local Theories of Causation and the A Posteriori Identification of the Causal Relation", *Erkenntnis* 48: 25–38.

Rogers, B. 1981: "Probabilistic Causality, Explanation, and Detection", *Synthese* 48: 201–223.

Rosen, D. 1978: "Discussion: In Defense of a Probabilistic Theory of Causality", *Philosophy of Science* 45: 604–613.

Rosenberg, A. 1992: "Causation, Probability and the Monarchy", *American Philosophical Quarterly* 29: 305–318.

Russell, B. 1913: "On the Notion of Cause", *Proceedings of the Aristotelian Society* 13: 1–26.

Russell, B. 1926: *Our Knowledge of the External World*. London: George Allen & Unwin.

Russell, B. 1948: *Human Knowledge*. New York: Simon and Schuster.

Sachs, R. 1972: "Time Reversal", *Science* 176: 587–597.

Sachs, R. 1987: *The Physics of Time Reversal*. Chicago: University of Chicago Press.

Salmon, W. 1970: "Statistical Explanation", in Colodny, R. (ed.), *The Nature and Function of Scientific Theories*, Pittsburgh: University of Pittsburgh Press, 173–231.

Salmon, W. 1971: *Statistical Explanation and Statistical Relevance*. Pittsburgh: University of Pittsburgh Press.

Salmon, W. 1978: "Why ask, 'Why?'?" *Proceedings of the American Philosophical Association* 51: 683–705.

Salmon, W. 1979: "Propensities: A Discussion Review", *Erkenntnis* 14: 183–216.

Salmon, W. 1980: "Probabilistic Causality", *Pacific Philosophical Quarterly* 61: 50–74.

Salmon, W. 1982a: "Causality: Production and Propagation", in Asquith, P. and Giere, R. (eds.), *PSA 1980*, East Lansing, Mich.: Philosophy of Science Association, 49–69.

Salmon, W. 1982b: "Further Reflections", in McLaughlin, R. (ed.), *What? Where? When? Why?*, Dordrecht: Reidel, 231–280.

Salmon, W. 1984: *Scientific Explanation and the Causal Structure of the World*. Princeton: Princeton University Press.

Salmon, W. 1985: "Conflicting Concepts of Scientific Explanation", *Journal of Philosophy* 82: 651–654.

Salmon, W. 1988a: "Dynamic Rationality: Propensity, Probability and Credence", in Fetzer, J. (ed.), *Probability and Causality: Essays in Honor of Wesley C. Salmon*, Dordrecht: Reidel, 3–40.

Salmon, W. 1988b: "Publications: An Annotated Bibliography", in Fetzer, J. (ed.), *Probability and Causality: Essays in Honor of Wesley C. Salmon*, Dordrecht: Reidel, 271–336.

Salmon, W. 1990: "Causal Propensities: Statistical Causality vs. Aleatory Causality", *Topoi* 9: 95–100.

Salmon, W. 1994: "Causality without Counterfactuals", *Philosophy of Science* 61: 297–312.

Salmon, W. 1997: "Causality and Explanation: A Reply to Two Critiques", *Philosophy of Science* 64: 461–477.

Salmon, W. 1998: *Causality and Explanation*. New York: Oxford University Press.

Salmon, W. and Wolters, G., eds. 1994: *Logic, Language and the Structure of Scientific Theories*. Pittsburgh: University of Pittsburgh Press.

Sayre, K. 1977: "Statistical Models of Causal Relations", *Philosophy of Science* 44: 203–214.

Shoemaker, S. 1984: *Identity, Cause and Mind*. Cambridge: Cambridge University Press.

Skyrms, B. 1980: *Causal Necessity*. New Haven: Yale University Press.

Smart, J. J. C. 1963: *Philosophy and Scientific Realism*. London: Routledge and Kegan Paul.

Sober, E. 1985: "Two Concepts of Cause", *PSA 1984* 2: 405–424.

Sober, E. 1987: "Explanation and Causation", *British Journal for the Philosophy of Science* 38: 243–257.

Sosa, E. and Tooley, M., eds. 1993: *Causation*. Oxford: Oxford University Press.

Strawson, G. 1989: *The Secret Connection*. Oxford: Clarendon.

Strawson, P. 1959: *Individuals*. London: University Paperbacks.

Suppes, P. 1970: *A Probabilistic Theory of Causality*. Amsterdam: North Holland.

Suppes, P. 1984: *Probabilistic Metaphysics*. Oxford: Blackwell.

Sutherland, R. 1983: "Bell's Theorem and Backwards-in-Time Causality", *International Journal of Theoretical Physics* 22: 377–384.

Sutherland, R. 1985: "A Corollary to Bell's Theorem", *Nuovo Cimento* 88B: 114–118.

Swinburne, R. 1981: *Space and Time*. 2nd ed., London: Macmillan.

Swoyer, C. 1984: "Causation and Identity", in French, P., Uehling, T., and Wettstein, H. (eds.), *Midwest Studies in Philosophy IX*, Minneapolis: University of Minnesota Press, 593–622.

Tooley, M. 1984: "Laws and Causal Relations", in French, P., Uehling, T., and Wettstein, H. (eds.), *Midwest Studies in Philosophy IX*, Minneapolis: University of Minnesota Press, 93–112.

Tooley, M. 1987: *Causation: A Realist Approach*. Oxford: Clarendon.

Tooley, M. 1990: "The Nature of Causation: A Singularist Account", in Copp, D. (ed.), *Canadian Philosophers Celebrating Twenty Years of the Canadian Journal of Philosophy*, Calgary: The University of Calgary Press, 271–322.

Tooley, M. 1993: "Introduction", in Sosa, E. and Tooley, M. (eds.), *Causation*, Oxford: Oxford University Press, 1–32.

Torretti, R. 1987: "Do Conjunctive Forks Always Point to a Common Cause?", *British Journal for the Philosophy of Science* 38: 384–387.

Tye, M. 1982: "A Causal Analysis of Seeing", *Philosophy and Phenomenological Review* 42: 311–325.

van Fraassen, B. 1989: *Laws and Symmetry*. Oxford: Clarendon.

von Wright, G. 1971: *Explanation and Understanding*. Ithaca: Cornell University Press.

von Wright, G. 1973: "On the Logic and Epistemology of the Causal Relation", in Suppes, P. (ed.), *Logic, Methodology and Philosophy of Science IV*, Amsterdam: North Holland, 293–312.

Whitrow, G. 1980: *The Natural Philosophy of Time*. 2nd ed., Oxford: Clarendon.

Winkler, K. 1991: "The New Hume", *The Philosophical Review* 100: 541–579.

Wright, J. 1983: *The Sceptical Realism of David Hume*. Manchester: Manchester University Press.

Index

Ackerman, F., 3
agency theory of causation *see*
    causation, manipulability theory
    of
Alston, W., 7
Anscombe, G., 5, 21
Aristotle, 53
Armstrong, D., vii, 3, 47, 52, 54, 77, 86,
    92, 105, 107, 126–8, 130–1, 144–5,
    168–70, 209
Arntzenius, F., 197
Aronson, J., 3–4, 6, 41–3, 44–6, 49,
    51–3, 55–7, 59–60, 64, 102, 111,
    117, 126
Arya, A., 24
'at-at' theory of causal transmission,
    67, 71

backwards-in-time causation, *see*
    causation, backwards-in-time
Beauchamp, T., 4, 5, 52, 59
Beebee, H., vii
Belinfante, H., 191
Bell, J., 25, 176–7, 180, 182, 184, 186,
    188–90, 192
Bennett, J., 52, 125
Berkovitch, J., vii
Bigelow, J., 1, 4–5, 9, 54, 111, 126
Blackburn, S., 21
Boltzmann, L., 196–7
Broughton, S., 21

Carrier, M., 114
Cartwright, N., 33, 148, 178
causal asymmetry, 57, 59–60, 177–80
    fork theory of, 179, 192–201, 204–6,
        208–9
    kaon theory of, 202–4

perspectival theory of, *see*
    subjective theory of causal
    asymmetry
subjective theory of, 179, 188–92
temporal theory of, 179, 187–8
causal connection, 170–5
causal lines, 62–6
causal processes, 64, 66–8, 72, 74–5,
    89, 91–4, 98–101
causality, *see* causation
causation
    backwards-in-time, 176–8, 180,
        183–92, 201, 204–10
    commonsense understanding of,
        2–3, 8–12, 46–8
    connotations of, 50–1
    Conserved Quantity theory of,
        89–122, 126, 130, 143, 147–9, 168,
        171–5
    counterfactual theory of, 12, 15,
        26–8, 126–7, 133, 142–4, 155–62
    and identity through time, 52–9, 91,
        101–9, 119–21
    indeterministic, 22–6, 33, 113
    invariant quantity theory of, 115–
        19
    manipulability (agency) theory of,
        12, 41–3, 49, 128, 188
    and negatives, 123–45
    physical, 12, 47
    probabilistic theory of, 30–40,
        150–9, 162–3
    regularity theory of, 14–26
    transference theory of, 41–61,
        109–11
chance-lowering causes, 33–40, 45–6,
    146–7, 151–4, 157–8, 161–2,
    164–7

Collingwood, R., 5, 128
common cause, principle of the, 68, 192
commonsense understanding of causation, *see* causation, commonsense understanding of
conceptual analysis, 1–3, 6–13
conjunctive forks, 69–71, 79–82, 85–6, 113, 117, 192–4, 198, 204–6, 209
connection, causal, *see* causal connection
connotations of causation, *see* causation, connotations of
conservation laws, 91, 94–8
Conserved Quantity theory of causation, *see* causation, Conserved Quantity theory of
Costa, M., 21
counterfactual theory of causation, *see* causation, counterfactual theory of
counterfactual theory of prevention and omission, 124, 132–45
Craig, E., 21
Cramer, J., 186, 203, 205

d'Espagnat, B., 184
Davis, W., 178
delaying and hastening, 141
despite defense, 38–40
Dieks, D., 6, 58, 111
disjunctive facts, 169–70, 174–5
Ducasse, C., 2, 5–7, 21
Dummett, M., 51

Earman, J., 6, 16, 24–5, 43, 189
Eells, E., 33–4, 39, 151–5, 178
Ehring, D., 53, 58, 105, 126, 130
Ellis, B., 54, 111, 124, 126, 132
Emmet, D., 4
empirical analysis, 1, 3–13
Enge, H., 23
entropy, 70, 195–8

Fair, D., 4, 6, 12, 41, 43–6, 56–9, 64, 102–3, 110–11, 124, 126
Fetzer, J., 71, 84–6
Feynman, R., 188
Flew, A., 5

Forge, J., 87, 114, 119
forks, *see* casual asymmetry, fork theory of

Ganeri, J., 135
Garrett, D., 15, 17
Gasking, D., 46, 49, 128, 164, 188
General Theory of Relativity, 96–7
genidentity, *see* causation and identity through time
Giere, R., 84
Glover. J., 125
Goldman, A., 125
Good, I., 30, 37
Goodman, N., 99
Grünbaum, A., 56, 200

Hall, N., 156
Hanna, J., 71, 76
Hart, H., 3, 6, 125
hastening, *see* delaying and hastening
Hausman, D., 126, 195, 207
Heathcote, A., 54, 111, 126
Hesslow, G., 156, 165
Hirsch, E., 105
Hitchcock, C., vii, 98
Honore, T., 3, 6, 125
Hume, D., 3–6, 14–29, 45, 84, 103–4, 187
Humean theory of causation, *see* causation, regularity theory of
Humphreys, P., 24, 34, 38, 76, 83, 85–6, 113

identity through time, *see* causation and identity through time
indeterministic causation, *see* causation, indeterministic
interactions, 67–74, 78–9, 82–3, 89–90
invariant quantities, *see* causation, invariant quantity theory of

Jackson, F., 1, 3, 48

kaon decay, *see* causal asymmetry, kaon theory of
Kistler, M., 57, 101
Kitcher, P., 70, 72, 74, 76, 78–9, 82–3, 89–90

Kolak, D., 105
Kripke, S., 4
Kvart, I., 162–3

Layzer, D., 203
Lee, T., 202
Leibniz, G., 52
Lewis, D., 9, 12, 14, 24, 26–8, 34–7,
    47–8, 102–3, 109, 126–7, 131, 133,
    142–4, 151, 155–61

Mackie, J., 2–3, 9, 25
Mackie, P., 141
manipulability theory of causation,
    see causation, manipulability
    theory of
mark method, 67, 74–9
Martin, R., 105
Mehlberg, H., 203
Mellor, D. H., 1, 6, 9, 20, 23, 38–40,
    50–1, 72, 81, 113, 128, 130–1,
    165–6
Menzies, P., vii, 12, 36–9, 48–50, 150–1,
    155–62, 177, 188
Miguel, H., vii
misconnections, 148–9, 167–8, 173–5
Mill, J. S., 5, 6
Millikan, R., 5
Molnar, G., vii
Mullany, N., 125
Musgrave, A., 107

Neander, K., 5
negatives in causation, see causation
    and negatives
Newton's First Law, 54, 63
Noordoff, P., 135, 163
Nozick, R., 105

O'Leary-Hawthorne, J., 159
omission, see causation and
    negatives

Papineau, D., 26, 38, 148, 195
Parfit, D., 105
Pargetter, R., 1, 4, 5, 9, 54, 111, 126
Pears, D., 21
persistence, see causation and identity
    through time

perspectivalism, see causal
    asymmetry, subjective theory of
physical causation, see causation,
    physical
Pitt, J., 66
Popper, K., 195
preemption, 155–7
prevention, see causation, and
    negatives
Pirce, H., vii, 10–11, 49–50, 128, 159,
    180, 188–91, 195–7, 205
probabilistic theory of causation, see
    causation, probabilistic theory of
pseudo processes, 64, 67, 75–9, 91–4,
    98–101

quantum mechanics
    Bell's theorem, 25, 180–7
    Copenhagen interpretation, 183
    backwards-in-time interpretation,
        183–6
    transactional account of, 186
quasi-dependence, 159–62
Quine, W., 55, 58, 91

Ramachandran, M., 127, 135
Ramsey-Lewis account of theoretical
    terms, 47–8
regularity theory of causation, see
    causation, regularity theory of
Reichenbach, H., 22, 30, 64–8, 78, 80,
    176, 192–4, 198–201, 209
relata of causation, 168–70
Reuger, A., 96–7
Rogers, B., 77, 83
Rosenberg, A., 4, 5, 26, 52, 59
Russell, B., 26, 29, 54, 62–7, 70

Sachs, R., 202–3
Salmon, W., vii, 4, 6, 24–5, 30, 33, 35,
    37, 62, 64–90, 92, 94, 98–9, 101–4,
    107–9, 111–22, 147–8, 167
Sayre, K., 84
Sellars, W., 169
Shoemaker, S., 105
Skyrms, B., 1, 90, 203
Smart, J., 180
Smith, N., 207
Sober, E., 33, 67, 69, 72, 150

Sosa, E., 66
Special Theory of Relativity, 64
Spinoza, B., 52
Spinozean disjunction, 55, 111
states of affairs, 168–70
Strawson, P., 3, 21–2
subjective theory of causal
    asymmetry, *see* causal asymmetry,
    subjective theory of
Suppes, P., 14, 20, 28–33, 38, 48, 126,
    128–9, 145
Sutherland, R., 185
Swinburne, R., 203
Swoyer, C., 105

Taylor, R., 104
Tooley, M., 6, 8, 20, 48, 52, 66, 71, 205,
    208–9
transference theory of causation, *see*
    causation, transference theory of
Tye, M., 125

Van Fraassen, B., 209
von Wright, G., 49, 128, 178, 188

Wheeler, J., 188
Whitrow, G., 203
Winkler, K., 21
Woodward, J., vii